Wissenschaftliche Reihe Fahrzeugtechnik Universität Stuttgart

Reihe herausgegeben von

André Casal Kulzer, Stuttgart, Germany

Hans-Christian Reuss, Stuttgart, Germany

Andreas Wagner, Stuttgart, Germany

Das Institut für Fahrzeugtechnik Stuttgart (IFS) an der Universität Stuttgart forscht interdisziplinär sowie technologieoffen an modernen und zukunftsorientierten Fahrzeugkonzepten. In enger Zusammenarbeit mit Partnern aus Industrie und Wissenschaft entstehen neue Lösungen für die Mobilität der Zukunft. Das Institut gliedert sich in drei spezialisierte Lehrstühle, die gemeinsam das gesamte Spektrum der Fahrzeugtechnik abdecken: Der **Lehrstuhl für Fahrzeugantriebssysteme** widmet sich der Forschung nachhaltiger Antriebslösungen für künftige Mobilitätskonzepte. Im Fokus stehen alternative, elektrische sowie hybride Antriebssysteme und deren Komponenten – einschließlich der Nutzung nachhaltiger Energieträger wie Wasserstoff, synthetischer Kraftstoffe und Batterien. Der **Lehrstuhl für Kraftfahrzeugmechatronik** beschäftigt sich mit vernetzten, intelligenten und adaptiven Fahrzeugen. Im Zentrum stehen Fragestellungen zum Automatisierten und Vernetzten Fahren, Diagnose, Ladetechnologien, verteilte Systeme sowie softwarebasierte Fahrzeugfunktionen. Der **Lehrstuhl für Kraftfahrwesen** erforscht die physikalischen Grundlagen der Auslegung zukünftiger Fahrzeugkonzepte. Im Mittelpunkt stehen die Bereiche Aerodynamik, Windkanaltechnik, Akustik/NVH, Fahrzeugdynamik, Reifenmanagement und Thermomanagement Gesamtfahrzeug. Das IFS verfügt über eine vielfältige und hochmoderne Forschungsinfrastruktur, die realitätsnahe Untersuchungen vom Einzelbauteil bis zum Gesamtfahrzeug ermöglicht. Besonders hervorzuheben sind der Multikonfigurations- und Antriebsprüfstand für komplexe Antriebskonzepte, der Stuttgarter Fahrsimulator zur Untersuchung menschlichen Fahrverhaltens, der Aeroakustik-Fahrzeugwindkanal für akustische und strömungstechnische Fragestellungen sowie der Thermowindkanal zur Analyse thermischer Prozesse im Gesamtfahrzeug. Die wissenschaftliche Reihe „Fahrzeugtechnik Universität Stuttgart" dokumentiert die im Rahmen von Promotionen am IFS entstandene Beiträge zur Mobilität der Zukunft und zeigt deren thematische Vielfalt, methodische Tiefe und Praxisrelevanz.

Reihe herausgegeben von

Prof. Dr.-Ing. André Casal Kulzer
Lehrstuhl für Fahrzeugantriebssysteme
Institut für Fahrzeugtechnik Stuttgart
Universität Stuttgart
Stuttgart, Deutschland

Prof. Dr.-Ing. Andreas Wagner
Lehrstuhl für Kraftfahrwesen
Institut für Fahrzeugtechnik Stuttgart
Universität Stuttgart
Stuttgart, Deutschland

Prof. Dr.-Ing. Hans-Christian Reuss
Lehrstuhl für Kraftfahrzeugmechatronik
Institut für Fahrzeugtechnik Stuttgart
Universität Stuttgart
Stuttgart, Deutschland

Mario Pipolo

3D-CFD Virtual Development of Post-Oxidation Strategies with Secondary Air Injection for Advanced Internal Combustion Engines

Springer Vieweg

Mario Pipolo
IFS, Institute of Automotive Engineering
University of Stuttgart
Stuttgart, Germany

Zugl.: Dissertation Universität Stuttgart, 2025 D93

ISSN 2567-0042 ISSN 2567-0352 (electronic)
Wissenschaftliche Reihe Fahrzeugtechnik Universität Stuttgart
ISBN 978-3-658-51342-9 ISBN 978-3-658-51343-6 (eBook)
https://doi.org/10.1007/978-3-658-51343-6

This Springer Vieweg imprint is published by the registered company Springer Fachmedien Wiesbaden GmbH, part of Springer Nature.
The registered company address is: Abraham-Lincoln-Str. 46, 65189 Wiesbaden, Germany

A tutti i miei colori.

Preface

This work was carried out during my time as a research associate at the Institute of Automotive Engineering (IFS) of the University of Stuttgart. This thesis marks the end of an intense, challenging, and deeply formative journey that would not have been possible without the support, guidance, and presence of many incredible people along the way.

I would like to express my sincere gratitude to Prof. Dr.-Ing. A. C. Kulzer for the opportunity to undertake this research, and for the continuous scientific support and guidance provided throughout the project. I am also deeply thankful to Prof. Yasuo Moriyoshi from Chiba University for accepting to be part of the doctoral committee and for the valuable insights and encouragement.

A special thank you goes to Dr.-Ing. Marco Chiodi, who has been an outstanding mentor and team leader. His constant support, enthusiasm, and technical advice have been fundamental for both my professional and personal development. I truly appreciated every discussion, challenge and word of encouragement shared along the way.

I was fortunate to share this experience with a team of brilliant and inspiring colleagues and friends — Francesco, Antonino, Edoardo, Cristian, Thomas, Robin, Peter and many others. Thank you for making work feel like a place where I could grow, belong, and enjoy the ride.

Outside the world of simulations and research, I've had the immense luck of walking this path with an amazing group of friends — Martina, Antonio, Nicole, Igor, Natalie, Carlo, Alfonso, Saverio and many more who made every step lighter. Thank you for turning ordinary moments into unforgettable memories.

I would like to dedicate this work to those who have always been by my side — my family, my sweet Dalila and my little Emily. Without your encouragement and presence, this achievement would not have been possible. I am deeply thankful to have you in my life.

Stuttgart   Mario Pipolo

Contents

Figures

Tables

Abbreviations

H_2	Hydrogen
CO_2	Carbon dioxide
CO	Carbon monoxide
H_2O	Water
N_2	Nitrogen
NO_x	Oxides of nitrogen
O_2	Oxigen
T_3	Exhaust gas temperature before turbine
p_3	Exhaust gas pressure before turbine
$p_{2.2}$	Intake pressure before cylinder
0D	Zero-dimensional
1D	One-dimensional
3D	Three-dimensional
a.FTDC	after firing top dead center
AFR	Air-fuel ratio
b.FTDC	before firing top dead center
BDC	Bottom dead center
BMEP	Break mean effective pressure
CA	Crank angle
CAD	Computer-aided design
CFD	Computational fluid dynamics
cm	Centimeter
CPU	Central processing unit
Cyl.	Cylinder
DI	Direct injection
DISI	Direct injection spark ignition
DNS	Direct numerical simulaion

ECU	Electronic control unit
EGR	Exhaust gas recirculation or residual burned gas of previous cylce
EOI	End of injection
EVO	Exhaust valve opening
FKFS	Forschungsinstitut für Kraftfahrwesen und Fahrzeugmotoren Stuttgart
FTDC	Firing top dead center
FVV	Forschungsvereinigung Verbrennungskraftmaschinen
GDI	Gasoline direct injection
GHG	Green house gases
HC	Unburned hydrocarbon
HCCI	Homogeneous charge compression ignition
HOV	Heat of vaporization
ICE	Internal combustion engines
IDT	Ignition delay time
IFS	Institut für Fahrzeugtechnik Stuttgart
IMEP	Indicated mean effective pressure
IP	Ignition point
IVC	Intake valve closure
kW	Kilowatt
LES	Large eddy simulation
LFS	Laminar flame speed
LHV	Lower heating value
MFB10-90	Combustion duration corresponding to 10-90% burned mass fraction
MFB50	50% Burned mass fraction
mm	Millimiter

PM	Particulate matter
PMAX	Maximum pressure
PNC	Particulate number concentration
Pox-rate	Post-oxidation rate
PRF	Primary reference fuel
RANS	Reynolds average navier stokes
RDE	Real driving emissions
rpm	Revolutions per minute
SAI	Secondary air injection
SI	Spark-ignition
SOI	Start of injection
TC up	Turbine upstream
TDC	Top dead center
TKE	Turbulent kinetic energy
TWC	Three way catalyst
WLTC	Worldwide harmonized light duty test cycle
WP	Real working process

Symbols

	Latin Letters	
A	Area	m^2
K	Wrinkling factor	-
S_L	Laminar flame speed	m/s
S_T	Turbulent flame speed	m/s
c_f	Long-range proces term in conservation equation	-
c_p	Specific heat at constant pressure	J/kgK
F	Extensive variable in conservation equation	-
f	Density / intensive variable in conservation equation	-
G	Gravitational energy	J
H_E	Exhaust enthalpy	J
h_f	Specific heat of formation	J/kg
H_I	Intake enthalpy	J
H_L	Leakage enthalpy	J
U	Internal energy	J
h_{tc}	Specific thermo-chemical enthalpy	J/kg
h	Specific enthalpy	J/kg
p	pressure	Pa
u	Specific internal energy	J/kg
v	Specific volume	m^3/kg
i	Stroke factor	-
$\vec{j}$	Diffusiv mass flux	kg/m^2s
k	Mass specific turbulent kinetic energy	m^2/s^2/kg
l	Characteristic length	m
M	Molar mass	kg/kmol
m	Mass	kg
m_B	Fuel mass	kg
m_C	Cylinder mass	kg

$\dot{m}$	Mass flow	kg/s
n	Engine speed	rpm
Oh	Ohnesorge number	-
P_e	Effective power	W
pme	Mean effective pressure	Pa
$\overline{\overline{P}}$	Second order stress tensor	N/m^2
$p\bar{\bar{I}}$	Momentum due to pressure	N/m^2
Q_B	Fuel heat release	J
Q_W	Wall heat transfer	J
R_s	Specific gas constant	J/kgK
Re	Reynolds number	-
s_f	Source or sink term in conservation equation	-
T	Temperature	K
t	Time	s
$\vec{V}$	Diffusiv fluid velocity	m/s
V	Volume	m^3
$\vec{v}$	Fluid velocity	m/s
V_{cyl}	Engine displacement	m^3
W	Work	J
w	Mass fraction	kg/kg
We	Weber number	-

	Greek Letters	
α	Spray jet angle	deg
ε	Geometric compressure ratio	-
η	Dynamic viscosity	Ns/m^2
η_f	Droplet dynamic viscosity	Ns/m^2
γ	Hollowcone spray angle	deg
α_{st}	Air to fuel ratio stoichiometric	-
λ	Air to fuel ratio	-
Ω	Fixed volume	m^3
$\Delta\Omega$	Surface of volume Ω	m^2
ω_i	Molar formation rate of species i	kmol/m^{3}s
φ	Crank angle	deg
φ	Spray azimuth angle	deg

$\vec{\Phi}_f$	Flux of the general density variable f in the conservation equations	-
$\overline{\overline{\Pi}}$	Momentum due to viscous forces	N/m^2
ρ	Density	kg/m^3
ρ_g	Density of the droplet surrounding gas	kg/m^3
ρ_i	Density of species i	kg/m^3
σ	Surface tension	N/m
σ_f	Droplet surface tension	N/m
Δt	Time step	s
τ_{ID}	Ignition delay time	s
τ	Characteristic droplet breakup time	s
θ	Spray zenith angle	deg

Abstract

The upcoming automotive regulations aim to accelerate the transition to near-zero-emission vehicles. With the introduction of Real Driving Emissions cycles and increasingly stringent environmental standards, the automotive industry is under pressure to develop effective strategies for limiting the environmental impact of conventional powertrains. Reducing fleet CO_2 levels presents significant challenges for internal combustion engine design, production, and certification. As a result, there is considerable interest in identifying and implementing innovative approaches to achieve these targets while optimizing costs and development time. A particularly critical aspect is the cold start phase, during which the three-way catalyst has not yet reached its activation temperature. The catalyst remains ineffective in this condition and cannot adequately convert harmful compounds. Therefore, accelerating the warm-up process is essential to comply with stricter regulatory limits.

In this context, the proposed work attempts to thoroughly investigate the potential of post-oxidation strategies with secondary air injection to reduce the light-off time of a conventional catalyst. This strategy involves injecting air into the exhaust manifold to oxidise unburned fuel from rich combustion within the cylinder. As the exhaust event of each cylinder occurs, hot exhaust gases are quickly discharged into the exhaust manifold, where fresh air is supplied through secondary air injectors located near the exhaust valves. The interaction between the hot gases and the fresh air initiates post-oxidation reactions, which release additional heat, thus increasing the temperature within the exhaust manifold. This rise in temperature creates a self-sustaining effect, ensuring that the chemical reactions continue throughout the exhaust stroke, further heating the catalyst and reducing the time needed to reach the light-off temperature. This approach can be implemented without significant modifications to the engine hardware or control system, making it a cost-effective option for retrofitting older engines or incorporating it into new designs. The investigation was conducted through both experimental and numerical methods, utilising test bench measurements and 3D-CFD simulations.

The experimental setup consists of a turbocharged, four-cylinder, direct injection, spark ignition engine and a low-inertia Horiba dynamometer to simulate driving conditions. The exhaust gas composition is measured with a gas analyser (Mexa-1600), and the pressure, temperature and lambda are collected at several points. A secondary air injection system was installed approximately 50 mm from the exhaust valve to ensure adequate oxygen for post-oxidation with a constant air flow. Even when the engine runs with excess fuel, the secondary air injection and subsequent oxidation of rich combustion products result in stoichiometric conditions before the catalyst, allowing it to operate within its optimum range. The oxidation reaction occurring in the exhaust manifold generates a significant amount of heat, increasing the enthalpy and temperature of the gases before the catalyst. This heat can be beneficial during cold starts and engine operating points requiring catalyst heating.

Post-oxidation phenomena are influenced by fundamental aspects such as mixing conditions, high temperature, availability of reactive combustion products, and time. Studying phenomena in the exhaust manifold requires 3D-CFD simulations, as it is challenging to obtain a visualisation of them through experimental techniques. The virtual engine test bench accurately describes the flow field and the chemical reactions within the numerical domain, allowing for the quantification of mixing and the chemical heat release rate. However, simulations of the exhaust system often lack reliable boundary conditions from the engine that consider gas composition and distribution. Therefore, the numerical investigations performed in the context of this study consist of two approaches. The first involves complete engine simulations using the 3D-CFD tool *QuickSim*, a fast-response simulation tool developed and continuously improved for over 20 years at FKFS (Forschungsinstitut für Kraftfahrwesen und Fahrzeugmotoren Stuttgart) and IFS (Institut für Fahrzeugtechnik Stuttgart), University of Stuttgart. Thanks to its ICE-adapted models, this tool can simulate entire engine models and consecutive operating cycles within a limited computational time. This is possible thanks to a combination of coarse meshes and self-developed internal combustion engine models, ensuring high predictability. By simulating the whole engine, a reduction of the boundary conditions complexity is possible, therefore reducing the influence of the boundary conditions assumptions on the final results. The computational domain expands from the air filter to the catalyst, also including a 0D model of the turbocharger.

This approach enhances the understanding of overall engine behaviour and the post-oxidation phenomena, including mixture inhomogeneities and cylinder imbalances.

Detailed boundary conditions are derived from the full engine model, tracking exhaust gas concentration and its distribution over time. This information is then used as input for a refined exhaust model that closely examines post-oxidation through a chemical reaction mechanism. This kind of simulation is carried out using the commercial 3D-CFD software *StarCCM+* and the domain focuses on the exhaust manifold from the exhaust valves to the turbine inlet, including the secondary air injectors. The chemical reaction mechanism has been chosen according to the experience gained in previous studies, and it allows for visualising the oxidation reactions and tracking the pollutants. The coupling of full engine simulation and detailed exhaust model enables an accurate study of post-oxidation, and it is significantly time efficient since the full engine model already considers the overall engine behaviour, combustion, and oxidation, utilising phenomenological models.

The 3D-CFD simulation methodology was successfully validated by applying it to engine operating points previously tested at the test bench, focusing on accurately replicating the measurement approach within the numerical model. Special attention was given to simulate the behaviour of a thermocouple to enable a more realistic comparison between the temperature profiles from the test bench and the simulation. At the test bench, a Mexa 1600 gas analyser operates with a constant mass flow rate to measure pollutant concentrations in exhaust gases. In contrast, the 3D-CFD simulation calculates species concentrations based on chemical reaction mechanisms, which may not fully capture the complexities present in the actual system. To accurately replicate the test bench methodology in the simulation, it was crucial to reproduce the sampling of species at a constant mass flow rate, similar to that of the Mexa 1600 system. The validation process involved the full engine simulation model and the detailed model of the exhaust manifold. Both experimental and numerical data were examined to determine if any of the strategies had advantages in terms of post-oxidation.

This research offers a comprehensive analysis of post-oxidation across various engine operating conditions by means of secondary air injection to understand better the factors that influence its effectiveness. A primary goal of this study is to investigate the conditions that facilitate the oxidation of unburned combustion products. To achieve this, the full engine model was employed, providing deep insights into mixture formation and cylinder-to-cylinder imbalances and the subsequent evolution of post-oxidation precursors. These findings are crucial for understanding how specific engine operating conditions impact post-oxidation efficiency.

The analysis then establishes a reference for post-oxidation effectiveness at different engine operating points by considering a refined model of the exhaust manifold. These case studies clarify how the post-oxidation process behaves across different operating conditions. The study identifies key aspects such as mixing conditions, availability of oxygen and rich combustion product, high temperature and sufficient residence time. These factors influence the post-oxidation efficiency and the amount of heat released. Once a solid baseline for post-oxidation is established, the research explores various techniques to improve the process. These strategies include increasing the temperature of secondary air injection, changing exhaust manifold design, and employing secondary fuel injection during different phases of the engine cycle. Each approach targets specific precursors of post-oxidation, such as temperature, mixing conditions, or the distribution of reactive species in the exhaust.

This study explores how different exhaust system configurations and fuel injection strategies affect the conditions that control post-oxidation in modern combustion engines. It focuses on the interaction between oxygen availability, temperature distribution, and mixing processes within the exhaust runners. Additionally, the research analyses how various manifold layouts and bypass concepts influence these factors. The investigation also considers how changes in fuel injection timing alter the chemical and thermal environment of the exhaust flow, affecting the reactivity of combustion products and the subsequent heat release downstream of the cylinder. The methods tested in this study offer valuable insights into enhancing post-oxidation, leading to better emissions control, increased engine efficiency, and more sustainable solutions for internal combustion engines.

In conclusion, a 3D turbocharger model was developed numerically better to track the flow field within the exhaust system. Since post-oxidation is highly dependent on mixing conditions, this improvement allows for a more accurate analysis of oxidation phenomena, particularly those occurring in the turbine or upstream of the catalyst. The 3D model of the turbocharger has been integrated into the entire engine extending the application of numerical analysis into areas that were previously accessible only through test bench experiments and one-dimensional simulations. The findings of this study contribute to the ongoing advancement of cleaner and advanced internal combustion engine technologies. Future research should focus on refining injection strategies, optimizing exhaust manifold designs, and integrating advanced numerical simulations with experimental validation to maximize post-oxidation efficiency.

Kurzfassung

Die zukünftige Gesetzgebung für Kraftfahrzeuge strebt eine Beschleunigung für den Übergang zu nahezu emissionsfreien Antrieben an. Angesichts der Einführung von RDE Zyklen (Real Driving Emissions) und immer strengeren Umweltstandards sieht sich die Automobilindustrie einem enormen Druck ausgesetzt wirksame Strategien zu entwickeln, um die Umweltauswirkungen konventioneller Antriebsstränge zu verringern. Die Reduktion der CO2-Emissionen innerhalb der Fahrzeugflotte stellt eine erhebliche Herausforderung für die Konstruktion, Produktion und Zertifizierung von Verbrennungsmotoren dar. Daher besteht ein großes Interesse daran, innovative Ansätze zu identifizieren, um die geforderten Ziele zu erreichen und gleichzeitig die Kosten und Entwicklungszeiten zu optimieren. Einen besonders kritischen Zustand stellt die Kaltstartphase von Verbrennungsmotoren dar, in welcher die Arbeitstemperatur des Dreiwegekatalysator noch nicht erreicht ist. In diesem Zustand arbeitet der Katalysator ineffizient und kann schädliche Abgasverbindungen nicht ausreichend umsetzen. Aus diesem Grund ist die Beschleunigung des Aufwärmprozesses ein entscheidender Faktor, um die strengen gesetzlichen Grenzwerte einhalten zu können.

Im Rahmen dieser Herausforderungen zielt die vorliegende Arbeit darauf ab, das Potenzial von Nachoxidationsstrategien, insbesondere der Sekundärlufteinblasung, zur Verkürzung des Aufheizzeitraums eines herkömmlichen Katalysators eingehend zu untersuchen. Bei dieser Strategie wird Luft in das Abgassystem eingeblasen, um den unverbrannten Kraftstoff aus einer fetten Verbrennung im Zylinder zu oxidieren. Während dem Auslasstakt jedes Zylinders, werden die heißen Abgase in den Abgaskrümmer abgeführt, wo durch Sekundärlufteinblasdüsen nahe der Auslassventile Frischluft zugeführt wird. Die Wechselwirkung zwischen den heißen Abgasen und der Frischluft löst Nachoxidationsreaktionen aus, die zusätzliche Wärme freisetzen und die Temperatur im Abgaskrümmer erhöhen. Diese Erhöhung der Temperatur führt zu einem selbstverstärkenden Effekt, der sicherstellt, dass die chemischen Reaktionen innerhalb des Abgasstroms fortgesetzt werden. Dadurch wird der Katalysator weiter erhitzt und die Zeit bis zum Erreichen der Light-Off Temperatur verkürzt. Dieser Ansatz kann ohne wesentliche Änderungen der Motorhardware und dem Steuersystem umgesetzt werden, was ihn zu einer kostengünstigen Option für die Nachrüstung älterer Motoren macht, sowie für neue Konstruktionen. Die vorliegende Untersuchung wurde sowohl experimentell als auch numerisch unter Verwendung von Prüfstands Messungen und 3D-CFD-Simulationen durchgeführt.

Die Versuchsanlage umfasst einen turboaufgeladenen Vier-Zylinder mit Direkteinspritzung sowie ein Horiba Motorprüfstand mit geringer Schwungmasse zur Simulation von Fahrbedingungen. Die Abgaszusammensetzung wird mit einem Gasanalysator (Mexa-1600) gemessen, wobei Druck, Temperatur und Lambda an mehreren Messpunkten aufgezeichnet werden. Ein Sekundärlufteinblasungssystem wurde in einem Abstand von etwa 50 mm zu den Auslassventilen installiert, um eine ausreichende Sauerstoffzufuhr für die Nachoxidation bei konstantem Luftstrom sicherzustellen. Auch wenn der Motor mit einem Kraftstoffüberschuss betrieben wird, führt die Sekundärlufteinblasung und die anschließende Oxidation der verbliebenden Verbrennungsprodukte zu stöchiometrischen Bedingungen vor dem Katalysator, sodass dieser im optimalen Bereich arbeiten kann. Die Oxidationsreaktion im Abgaskrümmer erzeugt eine beträchtliche Wärmemenge, welche Enthalpie und Temperatur der Gase vor dem Katalysator erhöht. Diese Wärmezufuhr kann beim Kaltstart und unter speziellen Betriebsbedingungen, welche eine Erwärmung des Katalysators erfordern, von Vorteil sein.

Die Nachoxidationsphänomene werden durch grundlegende Faktoren wie Mischungsbedingungen, hohe Temperaturen, die Verfügbarkeit reaktiver Verbrennungsprodukte und durch die Zeit beeinflusst. Die Untersuchung dieser Phänomene im Abgassystem erfordert den Einsatz von 3D-CFD-Simulationen, da eine experimentelle Visualisierung dieser Prozesse sehr schwierig ist. Der virtuelle Motorprüfstand liefert eine präzise Darstellung des Strömungsfeldes und der chemischen Reaktionen im numerischen Bereich, was eine Quantifizierung der Mischprozesse und der chemischen Wärmefreisetzungsraten ermöglicht. Für die Simulation des Abgassystems fehlen jedoch oft verlässliche Randbedingungen des Motors, welche die Gaszusammensetzung und dessen Verteilung berücksichtigen. Die numerischen Untersuchungen dieser Studie basieren daher aus zwei Ansätzen. Der erste Ansatz umfasst vollständige Motorsimulationen mit dem 3D-CFD-Tool *QuickSim*, einer Simulationssoftware, die seit mehr als 20 Jahren am FKFS (Forschungsinstitut für Kraftfahrwesen und Fahrzeugmotoren Stuttgart) und am IFS (Institut für Fahrzeugtechnik Stuttgart) der Universität Stuttgart entwickelt und kontinuierlich verbessert wird. Mit seinen speziell auf Verbrennungsmotoren abgestimmten Modellen ist dieses Tool in der Lage, komplette Motormodelle und aufeinanderfolgende Betriebszyklen innerhalb kurzer Rechenzeit zu simulieren. Dies wird durch eine Kombination aus groben Netzen und speziell für Verbrennungsmotoren optimierten Modellen ermöglicht, die ein hohes Maß an Vorhersagekraft gewährleisten. Durch die Simulation des gesamten Motorsystems können die Randbedingungen vereinfacht werden, was den Einfluss von Annahmen über die Randbedingungen auf die Endergebnisse verringert. Der Modellbereich erstreckt sich vom Luftfilter bis zum Katalysator, einschließlich eines 0D-Modells des Turboladers. Dieser Ansatz trägt dazu bei, das Verständnis des Gesamtverhaltens des Motors und der Phänomene, die die Nachoxidation beeinflussen, zu verbessern, wie etwa Mischungshomogenität und Zylindervarianz.

Detaillierte Randbedingungen werden aus dem Gesamtmotormodell abgeleitet, wobei die Abgaszusammensetzung und deren zeitliche Entwicklung verfolgt werden. Diese Informationen werden als Input für ein verfeinertes Abgasmodell genutzt, das die Nachoxidation durch einen chemischen Reaktionsmechanismus untersucht. Diese Art der Simulation wird mit der kommerziellen 3D-CFD-Software *StarCCM+* durchgeführt, wobei der Fokus auf dem Abgaskrümmer zwischen Auslassventil und Turbineneinlass liegt. Der chemische Reaktionsmechanismus wurde auf der Grundlage vorangegangener Studien gewählt und ermöglicht die Visualisierung der durch Oxidationsreaktionen freigesetzten Wärme sowie den Verlauf der Schadstoffe.

Die 3D-CFD-Simulationsmethodik wurde erfolgreich validiert, indem sie auf Motorbetriebszustände angewendet wurde, die zuvor am Prüfstand untersucht worden waren. Dabei lag ein besonderer Fokus darauf, den experimentellen Messansatz so realitätsnah wie möglich im numerischen Modell nachzubilden. Ein wesentlicher Bestandteil dieses Validierungsprozesses war die detaillierte Simulation des Verhaltens eines Thermoelements, um einen präziseren und physikalisch konsistenteren Vergleich der Temperaturverläufe zwischen Prüfstand und Simulation zu ermöglichen. Am Prüfstand wird ein Mexa-1600-

Gasanalysator eingesetzt, der mit einem konstanten Massenstrom arbeitet, um die Konzentration verschiedener Schadstoffe im Abgasstrom zu messen. Im Gegensatz dazu berechnet die 3D-CFD-Simulation die Konzentration chemischer Spezies auf Grundlage von Reaktionsmechanismen und stöchiometrischen Zusammenhängen, wodurch bestimmte Phänomene des realen Systems nur eingeschränkt erfasst werden können. Um die Prüfstandsmethodik möglichst genau zu reproduzieren, war es daher notwendig, auch im numerischen Modell ein Abtastverfahren zu implementieren, das die Probenahme der chemischen Spezies bei einem konstanten Massenstrom sicherstellt – analog zu jener des Mexa-1600-Systems. Der Validierungsprozess umfasste sowohl das vollständige Motormodell als auch das detaillierte Modell des Abgaskrümmers. Durch den Abgleich experimenteller und numerischer Daten konnte überprüft werden, inwieweit die Simulation die physikalischen und chemischen Vorgänge im Abgassystem korrekt abbildet. Darüber hinaus wurden beide Datensätze herangezogen, um zu bewerten, welche der untersuchten Strategien potenzielle Vorteile hinsichtlich der Nachoxidation bietet und wie konsistent diese Vorteile zwischen Simulation und Prüfstand beobachtet werden können.

Diese Forschungsarbeit liefert eine umfassende und vertiefte Analyse der Nachoxidation unter verschiedenen Motorbetriebsbedingungen, einschließlich der Sekundärlufteinblasung, und stellt damit einen wichtigen Beitrag zum Verständnis emissionsbildender Prozesse in modernen Verbrennungsmotoren dar. Ein zentrales Anliegen dieser Arbeit besteht darin, die physikalischen und chemischen Mechanismen zu charakterisieren, die die Oxidation unverbrannter Verbrennungsprodukte bestimmen, sowie jene Randbedingungen zu identifizieren, die diesen Prozess begünstigen oder hemmen. Zu diesem Zweck wurde ein vollständiges, mehrdimensionales Motormodell verwendet, das eine detaillierte Beschreibung der Gemischbildung, der Zylinder-individuellen Verteilungen und der Entstehung von Abgaszuständen ermöglicht. Die aus diesem Modell gewonnenen Erkenntnisse sind entscheidend, um nachzuvollziehen, wie spezifische Betriebsparameter die Effizienz der Nachoxidation beeinflussen und wie diese Parameter im Rahmen zukünftiger Motorstrategien gezielt gesteuert werden können.

Das im Rahmen der Arbeit entwickelte Analyseverfahren liefert eine verlässliche und reproduzierbare Grundlage zur Bewertung der Nachoxidation unter unterschiedlichen Last- und Drehzahlbereichen. Durch die Einbeziehung eines verfeinerten und realitätsnahen Modells des Abgaskrümmers wird eine noch präzisere Abbildung der strömungsmechanischen und thermischen Bedingungen ermöglicht, die den Oxidationsprozess maßgeblich prägen. Die durchgeführten Fallstudien verdeutlichen, wie sich der Nachoxidationsprozess in verschiedenen Betriebszuständen entwickelt, und erlauben eine systematische Einordnung der wesentlichen Einflussgrößen. Hierzu zählen insbesondere die Qualität der Strömungs- und Stoffmischung, die Verfügbarkeit von Sauerstoff sowie fetten Verbrennungsprodukten, das lokale Temperaturniveau und die effektive Verweilzeit im Abgassystem. Diese Faktoren wirken in komplexer Weise zusammen und bestimmen sowohl die Reaktionsgeschwindigkeit als auch die insgesamt freigesetzte Wärmemenge.

Auf Grundlage dieser Analyse wurden anschließend verschiedene Optimierungsmaßnahmen untersucht, deren Ziel es ist, die Rahmenbedingungen für eine effiziente Nachoxidation gezielt zu verbessern. Zu diesen Maßnahmen gehören die Erhöhung der Temperatur der Sekundärlufteinblasung, die geometrische Anpassung des Abgaskrümmers sowie der Einsatz einer sekundären Kraftstoffeinspritzung in unterschiedlichen Phasen des Motorzyklus. Jede dieser Strategien adressiert unterschiedliche Teilaspekte des Nachoxidationsprozesses, etwa die Verbesserung der thermischen Randbedingungen, die Optimierung der Strömungsführung oder die gezielte Steuerung der Konzentration reaktiver Spezies im Abgas. Die im Zuge dieser Untersuchungen gewonnenen Erkenntnisse liefern wertvolle Hinweise für die Auslegung künftiger Motorkonzepte, die auf eine erhöhte thermische Effizienz und eine weitergehende Reduktion von Emissionen abzielen.

Abschließend wurde ein detailliertes 3D-Turboladermodell entwickelt, um das Strömungsfeld und die damit einhergehenden Mischungsprozesse im Abgassystem noch präziser zu erfassen. Da die Nachoxidation in hohem Maße von lokalen Mischungsbedingungen abhängt, bietet die Erweiterung des Modellierungsrahmens um ein dreidimensionales Turboladermodell deutliche Vorteile gegenüber herkömmlichen vereinfachten Ansätzen. Insbesondere ermöglicht sie eine genauere Untersuchung der Oxidationsprozesse im Bereich des Turboladers sowie im Abschnitt unmittelbar vor dem Katalysator, wo kleinräumige Temperaturschwankungen und Strömungsinteraktionen eine zentrale Rolle spielen. Die erfolgreiche Integration des 3D-Turboladermodells in das Gesamtmotormodell erweitert den Anwendungsbereich numerischer Simulationen erheblich und erschließt Bereiche, die bislang primär durch Prüfstandmessungen oder eindimensionale Untersuchungen zugänglich waren. Die Erkenntnisse dieser Arbeit leisten somit einen wichtigen Beitrag zur Weiterentwicklung emissionsarmer, leistungsfähiger und zukunftsorientierter Verbrennungsmotoren. Darüber hinaus bieten sie eine solide Grundlage für zukünftige Forschungsarbeiten, die sich der Verfeinerung von Einspritzstrategien, der Optimierung von Abgaskrümmergeometrien und der Entwicklung erweiterter numerischer Simulationsmethoden widmen werden, um die Effizienz der Nachoxidation weiter zu maximieren.

1 Introduction

The upcoming vehicle emissions regulations aim to accelerate the transition to near-zero emission vehicles in response to the escalating climate crisis. The urgency of reducing greenhouse gas emissions drives the automotive industry to innovate rapidly to minimise the environmental impact of internal combustion engines (ICEs) by adopting renewable biofuels and advanced operating strategies while ensuring cost and time efficiency. One of the most challenging aspects is reducing emissions during a cold start when the aftertreatment system is not yet at its optimum operating temperature [33]. Therefore, emissions of unburned hydrocarbons (HC), carbon monoxide (CO), and nitrogen oxides (NO_x) are particularly high. Reducing the time it takes for the catalyst to reach its activation temperature is critical to meet the increasingly stringent emissions regulations and the Real Driving Emissions (RDE) cycles of modern emission standards [69]. In this context, the proposed work aims to thoroughly investigate the potential of post-oxidation phenomena in reducing the light-off time, i.e. the time required for the TWC to reach its minimum operating temperature (light-off temperature). The strategy involves injecting air into the exhaust manifold via secondary air injectors to oxidise unburned fuel from rich combustion within the cylinder. It can be implemented without significant modifications to the engine hardware or control system, making it an attractive option for retrofitting older engines or incorporating it into new designs. Even if the engine is running in excess of fuel, the secondary air injection and subsequent oxidation of the rich combustion products cause the engine to have a lambda before the catalyst around stoichiometric conditions, allowing the TWC to operate in its optimum range. The oxidation reaction in the exhaust manifold releases heat that increases the enthalpy and temperature of the gases before the TWC, which can be helpful during cold starts and catalyst-heating engine operating points. The investigation was conducted experimentally and numerically, with test bench measurements conduced at Chiba University and 3D-CFD (three-dimensional computational fluid dynamics) simulations performed at the University of Stuttgart. The collaboration between these two institutions has been fundamental in providing valuable insights into the research topic,

M. Pipolo, *3D-CFD Virtual Development of Post-Oxidation Strategies with Secondary Air Injection for Advanced Internal Combustion Engines*, Wissenschaftliche Reihe Fahrzeugtechnik Universität Stuttgart,
https://doi.org/10.1007/978-3-658-51343-6_1

combining expertise from both experimental and computational domains to advance the understanding of emission reduction technologies. This research will focus exclusively on the methodology and results within the virtual environment obtained during the FVV project 1456 "EmiRed Innovative RDE Engine-out Emission Reduction". Post-oxidation phenomena depend on fundamental aspects such as mixing conditions, high temperature, local mixing conditions, and time. Since it is tough to visualise these aspects experimentally, studying the phenomena occurring in the exhaust manifold and the above elements requires using 3D CFD. The 3D-CFD simulation enables a precise description of the flow field and the chemical reactions inside the domain; moreover, it permits quantifying the mixing and the chemical heat release rate.

Simulations of exhaust systems often lack of accurate and reliable boundary conditions, mainly due to the limitations of experimental measurements. Crank-angle resolved data on the composition and spatial distribution of the exhaust gases cannot be provided by the test bench. This lack of detailed and time-resolved information leads to gaps in the accurate definition of the input parameters for the simulations, which are essential to capture the dynamic behaviour and transient phenomena within the exhaust system. For this reason, the numerical investigation relies on two approaches: a full engine simulation, which provides a better understanding of the overall engine behaviour and the phenomena affecting the post-oxidation, such as the inhomogeneities in the mixture and the imbalance among the cylinders. Detailed boundary conditions from the full engine model are extrapolated considering the exhaust gas concentration over time and distribution. The information is given to a refined exhaust model, where a chemical reaction mechanism studies the post-oxidation in detail. The coupling of full engine simulation and detailed exhaust model enables an accurate study of post-oxidation, and it is significantly time efficient since the full engine model already considers the overall engine behaviour, combustion, and oxidation, utilising phenomenological models.

The complete engine simulations are carried out using *QuickSim* [13], a 3D-CFD fast response tool developed at FKFS and designed to simulate internal combustion engines. Thanks to its ICE-adapted models, *QuickSim* allows the simulation of complete engine models and consecutive operating cycles within a limited computational time. This is possible thanks to a combination of coarse meshes and self-developed internal combustion engine models, ensuring high predictability. By simulating the whole engine, a reduction of the boundary conditions complexity is possible, therefore reducing the influence of the boundary conditions assumptions on the final results. The computational domain expands from the air filter to the catalyst, also including a 0D model of the turbocharger. The second type of simulation is carried out using the commercial 3D CFD software *StarCCM+* [58] and focuses on the exhaust manifold from the exhaust valves to the turbine volute. This model used information from the complete engine 3D-CFD simulation as boundary conditions and is characterised by a finer mesh and the implementation of a chemical reaction mechanism to simulate the oxidation process of rich combustion gases inside the exhaust manifold. Detailed experimental data were used to validate the simulation results, demonstrating the reproducibility of the key post-oxidation aspects, including temperature, exhaust gas composition, pressure waves, air consumption, etc. In particular, it was necessary to reproduce the specific behaviour of the measurement methods numerically. Attention has been paid to replicating the behaviour of a thermocouple to provide a more realistic comparison of the test bench and simulation temperature profiles. In addition, the MEXA gas analyser was introduced into the simulation to give a constant sampling of combustion gases as in the experiment. Both experimental and numerical data were considered to understand whether any strategy had advantages in terms of post-oxidation. Experimentally, temperature trends between measurement points were examined. Numerically, in addition to the same analyses conducted at the bench, it was possible to quantify local mixing phenomena and the actual heat release due to oxidation reactions. The exhaust event of each cylinder triggers the response. The hot exhaust gases are pushed into the exhaust manifold at high speed through the fresh air injected from the secondary air injectors close to the exhaust valve. In these conditions, the temperature and the mixing happening at the interface between hot gases and fresh air start the post-oxidation reactions that release heat, increasing the temperature in the exhaust manifold. The higher temperature assures a self-sustaining effect that allows the chemical reactions to continue for the exhaust stroke.

Finally, a 3D turbocharger model was developed numerically to better track the flow field in the exhaust. Post-oxidation is highly dependent on the mixing conditions, and this improvement allows for a more accurate study of oxidation phenomena, such as those occurring in the turbine or upstream of the TWC. The 3D model of the turbocharger has been implemented in the entire engine domain. Integrating the turbocharger model allows a detailed energy balance analysis of the engine as the whole system under steady-state and transient operating conditions. As a result, a change in the operating state of the internal combustion engine will result in a corresponding change in the operating state of the turbocharger and vice versa. This capability extends the application of 3D-CFD into an area previously accessible only through test bench experiments and 1D (One-dimensional)simulations, providing virtual insight into complex phenomena such as turbo lag, turbo matching, or emission formation during transient load changes.

1.1 A new era for ICEs: challenges and opportunities

Internal combustion engines have been central to automotive propulsion for over a century, powering transportation and industrial applications. However, this era is undergoing transformative changes driven by environmental pressures, regulatory shifts, and the need for sustainable technologies. Despite the rise of electric vehicles (EVs) and alternative power sources, ICEs still present unique opportunities for innovation, particularly in reducing emissions through advanced combustion strategies, fuel alternatives, and emissions control technologies. International agreements and regional policies, such as the Paris Agreement [3] and the European Union (EU) Fit for 55 program [2], have significantly shaped the global drive towards sustainability and reduced carbon emissions. These frameworks set ambitious targets to reduce greenhouse gases and address climate change, considerably influencing vehicle and fuel technology development. The Paris Agreement, adopted in 2015 at COP21, is a legally binding international treaty under the United Nations Framework Convention on Climate Change (UNFCCC) with the primary goal of limiting global warming to well below 2 °C above pre-industrial levels, ideally closer to 1.5 °C. The agreement mandates substantial reductions in greenhouse gas emissions (GHG) across various sectors to reach this target. Each signatory nation must submit its own Nationally Determined Contributions (NDCs), outlining its emissions reduction goals and strategies. Signatories, including all EU member states, have pledged to review and update these commitments every five years to ensure progress toward achieving net-zero emissions by 2050. The agreement includes long-term temperature goals, adaptation and resilience measures, and a financial and technical support framework for developing countries. To reach its climate neutrality goal by 2050, the EU is taking action to reduce emissions from cars, as road transport accounts for one-fifth of the EU carbon dioxide (CO_2) emissions [1]. In alignment with the Paris Agreement goals, the EU introduced the Fit for 55 program in 2021, an ambitious policy framework designed to reduce EU GHG from cars by 55% by 2030, compared with 2021, to reach the goal of zero emissions from new cars and vans by 2035 [22]. This package of proposals includes stricter emissions standards for vehicles, a phase-out of new petrol and diesel cars by 2035, and support for low-emission technologies and sustainable fuels. For ICE technology, this program emphasises the need for innovation in emissions reduction strategies, including improved combustion efficiency, exhaust after-treatment systems, and the integration of alternative fuels like hydrogen and e-fuels, which offer lower lifecycle emissions. The regulatory package initially aimed to phase out the internal combustion engine in line with the 2035 neutrality target. However, this target faced strong opposition from some countries, e.g. Germany and Italy, due to their respective investments in synthetic and biofuels. In March 2023, the regulatory package was revised to include a specific section approving the use of CO2-neutral fuels, including e-fuels [47]. As reported in the European Environment Agency (EEA) web report [21], the share of transport emissions in the overall EU GHG emissions increased steadily from 16.5% in 1990 to 28.6% in 2019. In 2022, transport accounted for approximately 28.9% of all EU GHG emissions. In 2022, transport GHG emissions were 25.9% higher than in 1990. Figure 1.1 shows direct EU GHG emissions trends over time for different modes of transport between 1990 and 2022 [21]. Other categories of transportation included aviation, navigation, and rail. GHG emissions in the EU are expected to grow from 29% in 2022 to 31% in 2050. EU GHG emissions from road transport are almost entirely due to light- and heavy-duty vehicles; powered two-wheelers and other road transport modes are responsible for less than 1% of overall transport emissions. In 2022, heavy-duty vehicles represented 6% of total GHG emissions. GHG emissions from heavy-duty vehicles have increased by 29.3% since 1990. Furthermore, heavy-duty vehicles emitted approximately the same amount of GHG as the navigation and aviation sectors combined.

Overall, road transport emissions 2022 totalled 18% of all EU total GHG emissions. According to the projections, emissions from the road transport sector will reduce by 22.0% and 55.4%, respectively, compared to 2022.

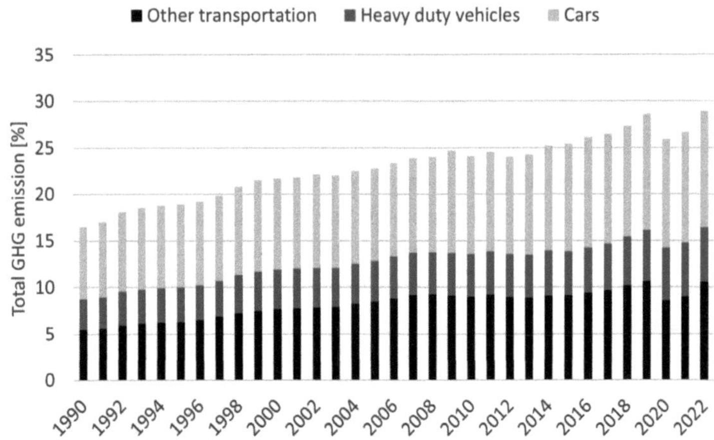

Figure 1.1: Shares of EU total GHG emissions.

The Paris Agreement and the Fit for 55 program have intensified regulatory studies on ICEs and highlighted the need for a balanced energy transition. In this transition, advanced ICE technologies can coexist with electric and hybrid powertrains, especially in regions with developing electric vehicle infrastructure. These regulations emphasise the enhancement of ICE efficiency and the development of sustainable fuels to ensure that ICEs can contribute meaningfully to a low-emission future in the short to medium term. The updated vehicle testing cycles, including the Worldwide Harmonized Light Vehicles Test Procedure (WLTP) and the Real Driving Emissions (RDE), are central to achieving these goals. These newer standards address limitations in the previous New European Driving Cycle (NEDC) by providing more realistic emissions assessments, fuel consumption, and electric vehicle range. The WLTP, with its standardised testing under variable driving patterns and conditions, provides a more accurate measurement of emissions and energy efficiency, aligning closely with the Fit for 55 targets by better reflecting real-world performance. Meanwhile, the

RDE test measures emissions directly on the road under dynamic driving conditions, thereby capturing emissions data that would typically be missed under laboratory tests alone. The RDE cycle is especially relevant for monitoring pollutants such as nitrogen oxides (NO_x) and particulate matter, contributing to air quality issues. The 2023 represents the third-year manufacturer CO_2 performance was determined using the WLTP instead of the New European Driving Cycle (NEDC), as reported in the research brief of the International Council on Clean Transportation [61]. The 2020 fleet-wide target of 95 g/km over the NEDC was converted to WLTP targets for 2021–2024. The conversion was performed for each manufacturer individually based on its average vehicle mass in 2020, where a higher mass leads to a higher CO_2 target and the ratio of WLTP to NEDC CO_2 emissions in 20210. Thus, each manufacturer is assigned a reference target for 2021–2024. The fleet average of all manufacturer-specific reference CO_2 targets is 118.1 g/km. The passenger car fleet CO_2 targets for 2025–2029 and 2030–2034 are determined by applying the respective 15% and 55% reduction rates to the 2021 baseline of 109.8 g/km. Applying these reduction rates yields a 93.6 g/km target for 2025 and a 49.5 g/km target for 2030 [20]. Based on the preliminary European Environment Agency (EEA) data, the sales-weighted fleet-average WLTP CO_2 emissions from new passenger cars in 2023 was 107 g/km, 1 g/km (1.5%) lower than in 2022 [61]. Figure 1.2 plots the average CO_2 emission values relative to the targets. Up to and including 2020, NEDC CO_2 values are reported; after 2020, WLTP values are presented. Before standards were introduced, fleet CO_2 emissions, on average, declined by 1.9 g/km per year from 2000 to 2007. After the first CO_2 standards were agreed upon in 2008, manufacturers outperformed the annual reduction rates required to meet the 2015 target of 130 g/km; instead of the required 3.6 g/yearly reduction km, average CO_2 emissions declined by 4.9 g/km per year. After 2015 targets were met, and without more stringent targets before 2020, average CO_2 emissions increased by 0.7 g/km per year. The 2020 target of 95 g/km over the NEDC included a phase-in provision and flexible compliance mechanisms, which led to a steep decline of 14 g/km from 2019 to 2020. Removing the phase-in provisions in 2021 required a further drop in fleet-average CO_2 emissions to meet the equivalent WLTP target. After this substantial reduction in 2021, manufacturers faced no pressure to further reduce CO_2 emissions until the next round of targets in 2025–2029, so 2022 and 2023 reductions occurred slowly.

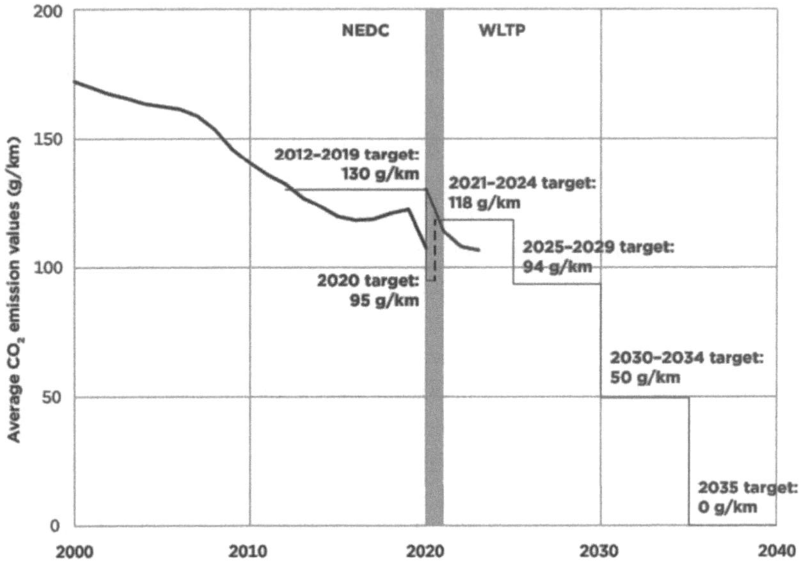

Figure 1.2: NEDC and WLTP CO_2 emission values and future targets.

A life-cycle assessment (LCA) is essential to identify the most suitable approach for decarbonising personal transportation. Alternative analyses, such as well-to-wheel evaluations, simplify the process and may unintentionally shift greenhouse gas (GHG) emissions to other stages of the production chain, thus failing to address the issue comprehensively. Since GHG emissions are a global challenge, moving emissions to other regions or stages, such as vehicle manufacturing or end-of-life management, without accounting for them does not resolve the problem. LCA is a formal and scientific methodology used to assess the environmental impacts of a product's life stages. LCA provides a systematic approach to measuring the ecological footprint, covering energy consumption, emissions, and other resources used and waste generated. LCA typically includes the following phases: goal and scope definition, inventory analysis, impact assessment, and interpretation. For the automotive industry, this can generally be divided into three main stages: production, usage, and end-of-life. Some studies [18] have shown that the GHG reduction potential of battery electric vehicles (BEVs) when considering the entire vehicle life-cycle, depends significantly on the energy mix used for electricity generation. The production of a BEV, which requires substantial energy, particularly for the battery manufacturing process, generates 80% more CO_2 equivalent emissions than conventional internal combustion engine vehicles (ICEVs), such as a gasoline-powered car. Mining raw materials such as lithium, cobalt, and nickel. crucial for EV batteries, can have significant environmental and social impacts. Additionally, the energy used in battery manufacturing, which often relies on conventional, non-renewable energy sources, adds considerable carbon emissions to the overall environmental footprint [48]. However, during the use phase, the overall emissions are determined by the grid mix of electricity used to charge the vehicle. Conversely, if the electricity is primarily derived from fossil fuels like coal, BEVs may not show significant GHG benefits compared to ICEVs [18]. Furthermore, the production of batteries often leads to adverse environmental impacts, such as increased acidification, human toxicity, particulate matter, photochemical ozone formation and resource depletion [18, 67]. This underlines that ICEVs, especially gasoline-powered ones, still present a viable and potentially environmentally beneficial option, particularly for middle-class consumers, due to their lower initial costs than BEVs and hybrid electric vehicles (HEVs) [55]. The highest level of GHG emissions in an ICEV occurs during its operational phase, primarily due to the exhaust gases released during fuel combustion. One promising approach to mitigate these emissions is using CO2-neutral fuels, such as e-fuels, which can minimise GHG emissions while benefiting from the lower environmental impact of ICEV production.

Renewable-produced synthetic fuels, called e-fuels, could be crucial in transitioning to low-carbon transportation. Among the various types of e-fuels, the most promising are liquid fuels, which can be synthesised via the methanol-to-gasoline process. These fuels rely on water and CO_2 captured from the atmosphere as raw materials [66]. The critical factor for the efficacy of e-fuels in reducing GHG emissions is the use of renewable energy during their production and the direct capture of CO_2 from the atmosphere [32]. In this context, ICEVs fuelled by e-fuels present a cost-effective, resource-efficient solution for achieving carbon-neutral transportation [46]. Moreover, e-fuels offer the advantage of being compatible with existing internal combustion engines without requiring modifications, which allows for a rapid and cost-effective conversion of current ICEVs into CO2-neutral vehicles [70]. Additionally, the existing fuel stations and pipeline infrastructure can be adapted to accommodate both e-fuels and biofuels without significant changes. E-fuels can be precisely engineered using a limited number of chemical species, which allows for optimal combustion characteristics and the reduction of harmful emissions from ICEs [5]. This enables the design of more efficient future engines and helps reduce emissions from current ICE vehicles [36]. Given these factors, ICEs running on liquid e-fuels could play a vital role in the future of transportation, not only for personal vehicles but also for industrial heavy-duty applications. It is essential to emphasise that, in the future, each technology should be applied where it is most suited to avoid pushing costly technologies that do not offer significant environmental benefits. For instance, BEVs are an excellent option for urban environments, where they reduce local emissions, but their environmental impact increases as the battery grows [67]. ICEVs fuelled by synthetic fuels are likely more environmentally sustainable for long-distance transportation.

1.2 The role of post-oxidation in emission reduction

The necessity of a remarkable reduction of CO_2 emissions, and specific competition with alternative powertrain concepts have begun the discussion about the future of internal combustion engines. The engine in the past has remained the same for many years, as the initial goal was to maximise performance. Since the 1990s, EURO regulations have started, and the focus has shifted towards pollutant emissions. To limit the environmental impact of internal combustion engines, the automotive industry must comply with increasingly stringent limits set by low-emissions regulations. The regulation extended the test method by considering the cold start emission measurement in the overall exhaust emissions of light passenger and commercial vehicles. In their study, Hu et al. [33] demonstrated that during cold starts, whether idle or driving, the Three-Way Catalyst (TWC) takes several minutes to reach its optimal temperature and become effective. Three-way catalysts are an effective after-treatment system to reduce CO, HC and NO_x emissions by converting them to CO_2, N_2 and H_2O. However, the catalytic conversion only starts after the TWC reaches its light-off temperature, typically between 250-300°C. Before this light-off temperature, the catalyst cannot effectively reduce emissions from the engine. Therefore, vehicle cold starts represent the primary source of air pollutants during the homologation cycle, including unburned hydrocarbons, carbon monoxide, oxides of nitrogen, and particulate matter [17, 27, 56, 68]. Reducing the time needed to reach the catalyst light-off temperature is crucial for the increasingly stringent emissions regulations. Many solutions have been proposed to support and induce significant catalytic activity in the early stages of engine operation and when the temperature of the equipment is below the light-off temperature. Active thermal management strategies rely on external devices and specific catalyst heat-up technologies to promote the warm-up of the catalytic converters. The adoption of burner-like systems and electrically heated catalysts has been identified as a promising solution, even though it is associated with an increased system complexity [8, 19, 31, 52]. As alternatives, in-cylinder and engine control strategies are typically exploited: late-combustion, idle speed increase, wastegate valve control and cylinder deactivation are among the most common solutions [25].

Kumar et al. [37] investigated scavenging-based post-oxidation phenomena with variable valve timing (VVT) and ignition timing tuning and found that the post-oxidation region was limited at long valve overlap. In addition, heterogeneity appeared in the exhaust manifold at long valve overlap, limiting effective post-oxidation. The effects of injection and ignition timing on post-oxidation were also investigated in [39]. It was noted that turbocharger speed increased to some extent as post-oxidation improved with advancing injection timing and decreased as the rate of fuel loss to the exhaust manifold decreased with further delay in injection timing. It was also noted that the post-oxidation key parameters indicate that the phenomena are highly dependent on appropriate temperature in the exhaust manifold, O2 availability, mixing condition and sufficient reaction time [39]. Improvement of the mixture in the exhaust manifold was also studied [38]. A bypass adapter was installed in the exhaust port, and improved mixing levels could improve post-oxidation and restore some of the thermal efficiency. H2 formation and its contribution to post-oxidation were also highlighted in [40], it was found that H2 is a crucial parameter for effective post-oxidation and that H2 emissions are higher in the rich combustion mode, which promotes post-oxidation. Most of the enthalpy generated in the exhaust manifold is due to the reaction between H_2 and CO. The same observations have been made in [38], highlighting the difficulties in determining the amount of post-oxidation and identifying the factors that affect it. Therefore, the investigation must rely on coupling the test bench and 3D-CFD simulations. Scavenging can be a challenging process to manage and may impact the engine's operation. 1D/3D models have also been developed, and post-oxidation phenomena with scavenging mechanisms have been studied [54, 62].

A secondary air system can inject fresh air near the exhaust valves. According to various literature sources, such as Lee D. et al. [42], secondary air injection (SAI) is a promising strategy to reduce *HC* emissions by 46% to 88% and *CO* emissions by 37% to 93% compared to operations without secondary air injection. In a sustainable post-oxidation reaction, factors such as mixture formation, temperature, mixing condition, and availability of rich combustion product in the exhaust manifold play a decisive role [40]. Serrano D. et al. [57] investigated catalyst light-off performance using SAI and found out that *HC* emissions and TWC light-off could be minimised under cold and warm conditions. Simulation and experimental validation with SAI were also conducted. The results show how SAI can provide significant benefits in terms of polluted emission reduction due to large oxygen availability and reduced catalyst light-off time [49]. Borland M. et al. [10] found out the location and amount of SAI, the exhaust manifold design, and ignition timing. They discovered that proper design and optimisation of SAI can improve catalyst light-off and *HC* reduction. Pritchard, J. et al. [53] investigated the SAI mechanism for the oxidation of particulate matter and found that its concentration could be reduced by up to 80%. However, Koehlen et al. [35] pointed out the knowledge of the precise conditions in the exhaust manifold (e.g. flow distribution, temperature, air and exhaust gas concentration) and cyclic fluctuations are decisive in enabling efficient post-oxidation, demonstrating the complexity of post-oxidation applications. One of the advantages of post-oxidation is that it does not require additional external energy from the vehicle elctrical system electricity to heat the TWC, but exploits instead the heat release rate of the reaction of rich combustion products with fresh air to warm the exhaust system. This can be implemented without the need for expensive systems. This factor will be important in the future because the use of hybrid systems will, on the one hand, require more frequent and faster heating of the TWC and, on the other hand, to keep the total cost of the car low, the cost of the engine will have to be limited.

1.3 Problem statement and research motivation

The automotive industry is under increasing pressure to improve the environmental performance of internal combustion engines due to the growing urgency to reduce GHG emissions and meet international climate targets, such as those outlined in the Paris Agreement [3]. The European Union's Fit for 55 initiative and new emission standards such as Euro 7 reinforce this demand by setting stringent CO_2 reduction targets and deadlines for achieving net-zero emissions [2, 22]. As a result, ICE technology must evolve rapidly to ensure compliance while maintaining performance and cost-effectiveness. A critical area in this evolution is addressing cold start emissions when engine catalysts have not reached optimum temperatures, resulting in higher unburned hydrocarbon, carbon monoxide and nitrogen oxide emissions [33, 69]. Moreover, the downsizing of direct injection spark ignition (DISI) engines using a turbocharger delays the torque build-up during transient operation (a phenomenon known as turbo-lag). Introducing RDE measurements increases the need for improved transient engine behaviour while reducing emissions. These challenges have stimulated interest in advanced post-oxidation strategies, where secondary air injection in the exhaust manifold promotes oxidation of unburned fuel, accelerates catalyst warm-up and reduces turbo-lag. However, effective control of post-oxidation requires accurate simulation tools that capture the complex physics of air-fuel interaction, turbulence and heat transfer in the exhaust system [49, 57]. The complexity of designing engines capable of meeting stringent emissions standards involves accurately modelling and predicting the physical phenomena associated, such as spray formation, combustion dynamics, and exhaust flow behaviour [20, 59]. 3D-CFD simulations have become an indispensable tool in this area, offering detailed insights into the fluid mechanics and thermodynamic processes within the engine. CFD models provide a cost-effective alternative to experimental testing but also present challenges, such as sensitivity to setup parameters, calibration dependencies, and high computational demands [50]. A dual-simulation approach combining full-engine models with detailed exhaust manifold simulations has been proposed to address these complexities. *QuickSim* enables full-engine simulations to capture overall engine dynamics, providing essential boundary conditions for exhaust modelling, such as gas composition, temperature, and pressure [58]. Furthermore, employing a refined mesh and incorporating a chemical reaction mechanism facilitates a comprehensive investigation of post-oxidation reactions and their impact on catalyst performance. As a result, combining full-engine and exhaust manifold simulations allows a more complete and efficient analysis, addressing the limitations of traditional simulation approaches and enabling the practical design of post-oxidation strategies.

Post-oxidation strategies, such as secondary air injection, offer a promising solution by injecting fresh air into the exhaust manifold, where it reacts with unburned hydrocarbons and CO, effectively increasing exhaust temperatures and enthalpy, advancing catalyst light-off and reducing turbo-lag [42, 49]. However, successful post-oxidation relies on achieving specific conditions, including adequate O_2 availability, sufficient exhaust gas temperature, and proper mixing of air and exhaust gases in the manifold. Therefore, advanced simulation models capable of representing the turbulence, temperature gradients, and chemical reactions within the exhaust system are essential to optimising SAI strategies. The strategy has been tested on a four-cylinder, turbocharged, direct injection, spark-ignition engine. Previous investigations of post-oxidation [54, 62, 63] already pointed out the difficulty of determining the amount of post-oxidation, as well as the most relevant influences. Therefore, the investigation must rely on a coupling of test-bench and 3D-CFD investigations. In the following, the procedure used to investigate post-oxidation employing SAI is presented with particular attention to 3D-CFD modelling. At first, post-oxidation was analysed at the test bench, where detailed measurements were performed. Afterwards, these measurements have been used to validate two different kinds of 3D-CFD simulations: full engine simulations with the 3D-CFD tool *QuickSim* [13], a detailed model of the exhaust manifold with chemical reaction mechanism and fine mesh.

At the end of this introduction, the reader can expect a detailed exploration of several key aspects of this research. Chapter 2 provides a comprehensive background to 3D-CFD simulations, covering the theoretical concepts and models used throughout the study. In Chapter 3, the focus shifts to the fundamental principles of chemical kinetics and thermodynamics, which are essential to understanding the post-oxidation process, how it occurs and why it occurs. This chapter also presents an in-depth discussion of the simulation of post-oxidation phenomena in a three-dimensional virtual environment, considering various reaction systems. It also provides an overview of the well-known post-oxidation strategies, as well as those investigated in this research. Chapter 4 presents a detailed description of the methodology used to carry out the analysis. This chapter includes a brief description of the experimental setup and virtual environment used to study the post-oxidation phenomena. The virtual environment consists of two models: one of the complete engine and one of the exhaust system, the results of which are presented in chapters 5 and 6 respectively. Chapter 7 provides an in-depth discussion of the 3D turbocharger model developed in the course of this research. It also compares the advantages and disadvantages of this approach with traditional methods. Finally, the conclusions and outlook summarise the findings and suggest future directions for research in this area.

2 Fundamentals of 3D-CFD Simulations for ICEs

Understanding internal combustion engines physical and chemical processes is fundamental to enhancing performance, improving fuel efficiency, and reducing emissions. The design and optimisation of ICEs involve analysing intricate phenomena such as air-fuel mixing, turbulent combustion, and pollutant formation. Computational Fluid Dynamics plays a central role in achieving this level of detailed analysis. Modern engine development relies on three simulation tools: Real Working Process (RWP) Calculation, 1D-CFD simulation, and 3D-CFD simulation [13]. Together, these tools form the Virtual Engine Development framework, which addresses different levels of complexity. This framework enables engineers to design engines that meet the demands of evolving technologies and increasingly strict emissions regulations. Chapter 2 examines the various methods and tools used to simulate and analyse the behaviour of internal combustion engines. It begins by exploring the Real Working Process, a fundamental approach to engine analysis that simplifies the complex interactions within an engine into global thermodynamic parameters. RWP provides an initial framework for understanding engine performance by assessing energy balance, gas exchange, and overall efficiency. Although it lacks the spatial resolution of more advanced techniques, RWP is a critical starting point, providing insights that inform subsequent simulation stages. The chapter then introduces 1D-CFD, which enhances the RWP by adding spatial resolution along the primary flow direction. This approach is convenient for studying intake and exhaust systems, turbocharger behaviour and transient operating conditions. By dividing the engine flow paths into discrete segments, 1D-CFD enables more detailed analysis while maintaining computational efficiency. Despite its limitations in capturing multi-dimensional phenomena, it remains a valuable tool in the iterative design and optimisation of engine systems. The chapter then moves on to 3D-CFD simulations, the most sophisticated and detailed method of studying engine behaviour. These simulations solve the Navier-Stokes equations over finely discretised domains, providing a comprehensive view of the interactions between fluid dynamics, turbulence, combustion and heat transfer.

M. Pipolo, *3D-CFD Virtual Development of Post-Oxidation Strategies with Secondary Air Injection for Advanced Internal Combustion Engines*, Wissenschaftliche Reihe Fahrzeugtechnik Universität Stuttgart,
https://doi.org/10.1007/978-3-658-51343-6_2

The theoretical basis of 3D CFD is discussed, including its ability to capture complex phenomena that more straightforward methods cannot address. Special attention is given to the conservation equations that form the basis of these simulations and to the challenges of modelling turbulence, an essential aspect of fluid dynamics in combustion engines. Finally, the chapter considers the role of turbochargers, critical components for improving engine efficiency and power output. This section explores the principles of turbocharging, its integration into modern engines, and the use of CFD tools to optimise its design and operation.

2.1 The real working-process analysis

The Real Working Process represents the foundational step in analysing the thermodynamic behaviour of an internal combustion engine. This approach is centred on solving key equations, including the first law of thermodynamics eq. 2.1, mass conservation eq. 2.3, and thermal equation of state eq. 2.4, which describe the basic principles governing engine operation [7, 45, 60].

- **First law of thermodynamics (Energy conservation)**
  The conservation of energy principle, expressed mathematically in eq. 2.1, represents the balance of energy contributions within the system. It accounts for the heat released by the fuel Q_B, the heat lost to the walls Q_W, the enthalpy of the intake air H_I, the enthalpy of the exhaust gases H_E, the enthalpy lost through leakage H_L, and the work done by the engine W. This equation asserts that, in an isolated system, the total energy remains constant, with energy transformations and transfers between these components maintaining the overall balance.

$$Q_B + Q_W + H_I + H_E + H_L + W = 0 \qquad \text{eq. 2.1}$$

  The differential form of eq. 2.1 represents the rate of change of internal energy U over time (eq. 2.2). By introducing the crank angle derivative, denoted as $\mathrm{d}\varphi$, this formulation enables a crank-angle-specific analysis of the energy balance throughout the engine cycle. This approach provides detailed insights into how energy transitions and transformations occur during each cycle phase, allowing for a more resolved and precise evaluation of the system thermodynamic behaviour.

$$\frac{dQ_B}{d\varphi}+\frac{dQ_W}{d\varphi}+\frac{dH_I}{d\varphi}+\frac{dH_E}{d\varphi}+\frac{dW}{d\varphi}+\frac{dH_L}{d\varphi}=\frac{dU}{d\varphi} \quad \text{eq. 2.2}$$

- **Conservation of mass**
  The variation in cylinder mass m_C can be expressed in eq. 2.3. It is influenced by several factors: the mass of exhaust gases m_E and leakage mass m_L leaving the cylinder due to blow-by, as well as the inflow of fresh charge m_I. For engines with direct injection, the injected fuel mass m_B also contributes to the overall balance of cylinder mass.

$$\frac{m_C}{\varphi}=\frac{m_I}{\varphi}+\frac{m_E}{\varphi}+\frac{m_L}{\varphi}+[\frac{m_B}{\varphi}]_{DI} \quad \text{eq. 2.3}$$

- **Thermal equation of state**
  The ideal gas law describes the relationship between the thermal state variables of an ideal gas gas, as shown in equation eq. 2.4. This equation connects the pressure p, volume V, mass m, specific gas constant R_s, and temperature T of the gas. Specifically, it provides a fundamental basis for understanding how these variables interact under ideal behaviour, where intermolecular forces and molecular volume are negligible. For a fixed amount of gas, this relationship explains how changes in one variable, such as pressure or temperature, affect the others. For instance, increasing the temperature T at constant volume V will result in a proportional increase in pressure p, reflecting the direct correlation dictated by the equation.

$$p\cdot V=m\cdot R_s\cdot T \quad \text{eq. 2.4}$$

2.2 1D-CFD simulation

1D-CFD simulations extend the fundamental analysis of the Real Working Process by introducing spatial resolution along the primary flow direction within the engine. In this approach, the engine domain, including intake and exhaust regions, is divided into discrete sub-volumes. Each sub-volume conservation equations for mass, momentum, and energy are solved iteratively, providing a time-resolved depiction of the fluid dynamics and thermodynamic processes [7, 13]. This methodology allows engineers to model the entire flow path through the engine, including intake and exhaust systems, turbochargers and other critical components. Specifict models can be incorporated for cylinder heat transfer, crankshaft friction and even temperature changes caused by thermocouple placement. For combustion, 1D-CFD uses phenomenological or quasi-dimensional models to account for factors such as combustion chamber geometry, piston design and in-cylinder flow patterns such as tumble and swirl. These models improve predictive capabilities by allowing simulations to account for transient phenomena such as rapid power changes or speed variations [4, 16, 43]. 1D-CFD simulations are particularly effective for studying engine performance over broad operating conditions. They are often used to:

- Analyze charging systems and perform turbocharger matching for optimised performance.
- Design intake and exhaust manifolds to enhance airflow and reduce back pressure.
- Investigate transient engine behaviour, such as acceleration or deceleration events.
- Optimize valve timing to improve efficiency and emissions.
- Simulate complete driving cycles like RDE and WLTP, providing insights into engine behaviour under real-world conditions.

By integrating additional subsystems, such as cooling circuits and camshaft dynamics, 1D-CFD offers a versatile environment for engine development. Its efficiency and broad applicability make it a widely used tool in automotive design and testing [7, 23]. Despite its strengths, 1D-CFD has inherent limitations due to its simplified spatial resolution. It cannot capture detailed three-dimensional phenomena critical for fuel spray targeting, mixture formation and scavenging. These limitations are particularly pronounced in cases where local resolution is required, such as post-oxidation studies, where the inhomogeneity of the exhaust flow is critical. The coarse discretisation in 1D-CFD prevents accurate simulation of such phenomena, requiring complementary 3D CFD simulations for validation and refinement. To overcome these limitations, 1D CFD simulations are often used in combination with 3D CFD models. While 1D-CFD provides a fast and computationally efficient overview of engine behaviour, 3D-CFD adds the spatial detail required to analyse local interactions, such as post-oxidation and in-cylinder combustion dynamics. By combining these tools, engineers can balance accuracy and computational efficiency, addressing large-scale engine behaviour and complex localised phenomena.

2.3 3D-CFD simulation

Three-dimensional Computational Fluid Dynamics is a powerful tool for the numerical investigation of thermo-fluid-dynamic processes in complex systems, such as internal combustion engines. Unlike more straightforward simulation methods, 3D-CFD allows for an in-depth analysis of intricate behaviours within geometrically complex environments, including fluid motion, mixing, combustion, and heat transfer. The core of 3D-CFD simulations lies in solving partial differential equations that govern the conservation of mass, momentum, energy, and species concentration within a defined fluid domain. The computational domain is divided into finite volumes, often called cells, where these equations are solved iteratively. Different cell shapes may be used depending on the desired accuracy and computational resources, such as tetrahedral, hexahedral, or octahedral. This meshing process ensures that even highly irregular geometries can be accurately represented in the simulation domain. A common assumption in 3D-CFD is that fluid properties and thermodynamic states within each cell are homogeneous, with the cell centre representing the discretised volume. Time is similarly discretised into small intervals, or time steps, to account for transient phenomena, such as combustion events or flow turbulence [6, 12, 13, 24, 65]. However, the computational cost and time associated with 3D-CFD are significantly higher than those associated with simpler models, such as 1D-CFD or Real Working Process Analysis. This chapter explores the mathematical framework underlying 3D-CFD, focusing on the fundamental equations that govern fluid dynamics and heat transfer. The discussion then transitions to discretisation methods, which are vital for solving these equations numerically. Finally, the chapter addresses the importance of boundary conditions in ensuring the accuracy and realism of CFD simulations.

2.3.1 Fundamental equations

It is essential to establish the fundamental principles that dictate fluid dynamics to describe the behaviour of fluids and the equations governing their motion. The state of a fluid can be defined using its key properties: density, velocity, pressure, and either internal energy or temperature. These properties are functions of time and space, and specific physical laws govern their evolution. For a fluid in three-dimensional space, six independent equations are needed to describe its state fully: the continuity equation (mass conservation), three momentum equations, the energy equation, and an equation of state relating the thermodynamic variables pressure, density and temperature The Euler formulation serves as the foundation for 3D-CFD simulations, offering a systematic framework for describing the motion and behaviour of fluids at fixed points in space. Euler's equations describe the behaviour of an ideal fluid, assuming the absence of viscosity and thermal conduction. The method examines changes in fluid properties, such as velocity, pressure, density, and temperature, at specific spatial locations over time. This stationary perspective suits it, particularly for complex engineering applications, such as ICEs, where detailed interactions between fluid dynamics and thermodynamics occur within confined geometries. As reported in eq. 2.5, the general Euler formulation is derived from the conservation laws for an extensive variable $F(t)$ applied to a control volume.

$$\frac{\partial f}{\partial t} + div\left(\vec{\Phi}_f\right) = s_f + c_f \qquad \text{eq. 2.5}$$

where $f(\vec{x},t) = dF/dV$ represents the density or intensive variable of the extensive variable $F(t)$ in the cell at position $\vec{x}$. The conservation equation states that a change in time of the density $\partial f/\partial t$ can be caused by flux through the volume surface $\vec{\Phi}_f$, a source or sink s_f, or a long-range process c_f like radiation or gravitation. Such a general formulation can be applied to the different variables of the CFD problem, and conservation equations for mass, momentum, and energy can be derived.

- **Mass conservation equation**
  Considering the Euler Formulation (eq. 2.5), the mass conservation equation can be derived taking into account the mass as extensive variable $F(t)$ and the density term $f(\vec{x},t)$ is given by the mass density ρ. The mass conservation

equation (eq. 2.6) ensures that mass is neither created nor destroyed. Thus, Both source and long-range terms can be set to 0; it states that the rate of change of mass in a control volume is equal to the net flux of mass through its boundary. The flux density term is represented by the product of the local flow velocity $\vec{v}$ and the density ρ. In its differential form, it is:

$$\frac{\partial \rho}{\partial t} + div\,(\rho \vec{v}) = 0 \qquad \text{eq. 2.6}$$

If i species are present in the system, for example, air and fuel vapour, the mass m is replaced by the mass fraction $w_i = m_i/m$ of species i. The density of the species i gives the density term. Mass density can be written as the product of local flow velocity v_i composed of the mean flow velocity $\vec{v}_i$ and the diffusion velocity $\vec{V}_i$. If $\vec{j}_i$ is considered as the derived mass flux created by the diffusion velocity, eq. 2.6 can be written as eq. 2.7.

$$\vec{\Phi}_f = \rho_i \vec{v}_i = \rho_i(\vec{v}_i + \vec{V}_i) = \rho_i \vec{v}_i + \vec{j}_i \qquad \text{eq. 2.7}$$

As species can be transformed into each other, e.g. through combustion or other chemical reactions, a source term is required which can be written as the product of species molar mass M_i and species molar formation rate ω_i. The eq. 2.6 is therefore derived.

$$\frac{\partial \rho w_i}{\partial t} + div\,(\rho w_i \vec{v}) + div\,\vec{j}_i = M_i \omega_i \qquad \text{eq. 2.8}$$

- **Momentum conservation (Navier Stokes equation)**
  When addressing the conservation of momentum (eq eq. 2.9), the density term ρ in eq. 2.5 is replaced by the momentum density $\rho\vec{v}$. The flux term, representing the transport of momentum, comprises two primary contributions: the convective transport of momentum $\rho\vec{v} \otimes \vec{v}$ due to the fluid motion and the second-order stress tensor $\overline{\overline{P}}$. This stress tensor accounts for changes in momentum caused by two distinct mechanisms: the effects of viscosity within the fluid $\overline{\overline{\Pi}}$, which represent internal friction forces, and the pressure forces applied by the surrounding fluid $p\overline{\overline{I}}$.

$$\vec{\Phi}_f = \rho\vec{v} \otimes \vec{v} + \overline{\overline{P}} = \rho\vec{v} \otimes \vec{v} + p\overline{\overline{I}} + \overline{\overline{\Pi}} \qquad \text{eq. 2.9}$$

This formulation is fundamental in deriving the Navier-Stokes equations central to fluid dynamics. Taking into account the long-range term, given by gravitation contribution $\rho\vec{g}$, the momentum conservation equation can be written as follow (eq. 2.10):

$$\frac{\partial(\rho\vec{v})}{\partial t} + div(\rho\vec{v}\otimes\vec{v}) + div\overline{\overline{\Pi}} - grad\,p = \rho\vec{g} \qquad \text{eq. 2.10}$$

- **Energy conservation**
  The conservation equation of energy ensures that the total energy including internal energy u, kinetic energy $\frac{1}{2}|\vec{v}|^2$, gravitational potential energy G, and the heat of formation h_f of the species involved is equal to the total energy density ρe, as described in eq. 2.11. This principle ensures that all forms of energy are accounted for and conserved within the system.

$$\rho\left(u + \frac{1}{2}|\vec{v}|^2 + G + h_f\right) = \rho e \qquad \text{eq. 2.11}$$

Energy flux within the fluid can be expressed as the sum of several terms: the convective transport $\rho e\vec{v}$, energy transport resulting from pressure and shear stresses $\overline{\overline{P}}\vec{v}$, and heat conduction $\vec{j}_q$. In processes where the creation or destruction of energy is not possible, such as in internal combustion engine systems, the source term s_f can be considered negligible. However, the long-range term c_f may be included to account for effects such as magnetic fields or radiative energy transfer, represented by q_r. The thermo-chemical enthalpy h_{tc} further refines the description of energy content. It is composed of the thermal enthalpy h, representing heat-related energy, and the heat of formation h_f, which accounts for the energy associated with the formation of specific chemical species. Considering these assumptions, the energy equation assumes the form shown by eq. 2.12.

$$\frac{\partial(\rho h_{tc})}{\partial t} - \frac{\partial p}{\partial t} + div\left(\rho h_{tc}\vec{v} + \vec{j}_q\right) + \overline{\overline{P}} : grad(\vec{v}) - div(p\vec{v}) = q_r \qquad \text{eq. 2.12}$$

The thermo-chemical enthalpy h_{tc} is widely used in 3D-CFD simulations of internal combustion engines as a critical variable for characterising the energy content of the fluid. It combines two components: the thermal enthalpy h and the chemical contribution h_f, effectively capturing both heat-related and chemical energy aspects of the fluid (eq. 2.13).

$$h_{tc} = h + h_f \qquad \text{eq. 2.13}$$

2.3.2 Turbulence modelling

Turbulence modelling is a fundamental aspect of 3D-CFD simulations for ICEs, as the flow inside an engine operates predominantly in the turbulent regime. Turbulence introduces complex and non-stationary structures at varying scales, ranging from large vortices influenced by geometry to small-scale isotropic eddies. Accurately capturing these phenomena is critical for simulating engine behaviour, including combustion processes, mixture formation, and heat transfer, which are all heavily influenced by turbulent flows [9, 45, 65]. In turbulent flows, characterised by a characteristic length l, the fluid motion is marked by chaotic and unpredictable fluctuations in velocity, which in turn cause variations in properties like density, temperature, and mixture composition. These fluctuations arise primarily due to the shear stresses within the flow. These stresses form diverse sizes and shapes of vortices, continuously extracting energy from the primary flow. This energy is transferred to smaller vortices through a cascade process, ultimately dissipating as heat through viscous effects [6, 13]. A valuable metric for distinguishing flow regimes is the *Reynolds number*, a dimensionless quantity that quantifies the relative influence of inertial and viscous forces. By comparing these forces, the Reynolds number helps determine whether the flow exhibits a laminar or turbulent behaviour ($Re > 2.30$ for a flow in a pipe with a diameter l), as described in eq. 2.14.

$$Re = \frac{\rho|\vec{v}|l}{\mu} \qquad \text{eq. 2.14}$$

where ρ is the density, $|\vec{v}|$ is the velocity, l is the characteristic length and μ is the dynamic viscosity. Turbulent flows are characterised by two main length scales: the integral and Kolmogorov scales. The integral length scale l_l represents the largest turbulent eddies within the flow, the size of which is determined by the system geometry. These large vortices contain and transmit most of the flow kinetic energy. At the opposite end, the Kolmogorov scale l_k refers to the smallest vortices where turbulence becomes viscous dissipation. At this microscopic level, the kinetic energy of the smallest vortices is converted into internal energy and eventually dissipated as heat. The mechanism governing this process is the energy cascading associated with turbulent flows. Energy is introduced at large scales, where the largest eddies are formed. These eddies are unstable, breaking down into smaller and smaller vortices as energy is transferred through intermediate scales. This cascade continues until it reaches the Kolmogorov scale, where viscosity dissipates energy. This systematic redistribution of energy across scales defines the chaotic nature of turbulence. Accurately representing the energy cascade is essential for selecting appropriate turbulence models. Several approaches are available to model turbulence, each offering trade-offs between computational cost and physical accuracy. The most detailed method, *Direct Numerical Simulation (DNS)*, involves directly solving the Navier-Stokes equations without approximations or models for turbulence. This approach resolves all turbulence scales, including the smallest vortices dictated by the Kolmogorov length scale, which is proportional to the *Reynolds number*. While *DNS* provides unparalleled detail, it is computationally prohibitive for engine simulations due to the enormous number of grid points and time steps required, often making it impractical for high *Reynolds number* flows such as those found in ICEs. [9, 13, 45]. An intermediate approach is *Large Eddy Simulation (LES)*, which explicitly resolves the larger eddies while modelling the smaller, isotropic ones using subgrid-scale models. *LES* balances accuracy and computational cost, making it suitable for applications requiring detailed turbulence analysis, such as spray dynamics. However, *LES* requires highly refined computational meshes and is primarily used in research contexts rather than iterative engine design due to its significant computational demand [13]. *Reynolds-Averaged Navier-Stokes (RANS)* models are the most widely used for practical engineering applications. *RANS* relies on averaging the Navier-Stokes equations over time and decomposing the velocity field into mean and fluctuating components. This method introduces turbulence models, such as the widely applied $k-\varepsilon$ model, which uses additional transport equations to predict turbulent kinetic energy k and its dissipation rate ε. The $k-\varepsilon$ model has been extensively validated for confined flows, such as those in internal combustion engines, offering a reasonable compromise between computational efficiency and physical accuracy. *RANS* models significantly reduce computational effort compared to *DNS* and *LES*, enabling simulations across large computational domains and high *Reynolds number* flows typical of engine environments [6, 13].

2.3.3 Integrated turbocharger simulations

The increasing computational power and availability of more detailed numerical models allow the investigation of phenomena previously not considered, e.g. a detailed turbocharger simulation as an integrated part of 3D-CFD engine simulations. Introducing the turbocharger in the virtual engine test bench permits the application of boundary conditions matching the actual test bench conditions. This is particularly valuable in the early stages of engine development when no experimental data are available for calibration and validation of 1D and 3D simulation models. Different approaches are used to balance computational time and reliability of results [34], of which the following are of particular importance:

- 0D turbocharger model: Continuous data exchange is enabled between the 3D simulation and the 0D turbocharger model, with processed information, such as a new shaft speed, returned as global data in the form of an internal boundary condition. This approach ensures that the boundary conditions for full engine simulations are air at ambient pressure and temperature without specific three-dimensional flow assumptions, thereby reducing boundary condition dependency.
- 3D turbocharger model: this can be seen as a hybrid solution. At the intake, the boundary conditions are directly taken at the compressor outlet, while a turbine geometry is implemented at the exhaust. This allows a high level of detail at the exhaust side, especially for precise evaluation of the flow field to study local mixing phenomena.

The second approach is only viable if the simulation time remains within acceptable limits and the model demonstrates stability, particularly given the numerous pressure pulsations in the exhaust of a turbocharged engine that must be represented with precision. This methodology was developed and implemented by the author, providing a reliable framework for accurate simulation of turbocharged engines. Further details about this approach, its development, and its application will be presented in Chapter 7, where its contributions to the overall analysis are discussed in depth.

3 Post-Oxidation in Internal Combustion Engines

Numerous approaches have been proposed to enhance and trigger significant catalytic activity during the early stages of engine operation, mainly when the temperature of the exhaust system is below the so-called "light-off"threshold. Active thermal management techniques often rely on external devices and specialised catalyst heating technologies to accelerate the warm-up of TWC. Systems such as burner-like devices and electrically heated catalysts are promising, although they come with the trade-off of increased system complexity [8, 19, 31, 52]. As alternative methods, in-cylinder and engine control strategies are commonly employed, including late combustion, increased idle speed, wastegate valve control, and cylinder deactivation [25]. In this context, inducing post-oxidation reactions presents an effective way to speed up catalyst heating: rich in-cylinder combustion generates unburned by-products that can undergo further oxidation in the exhaust manifolds using specific strategies. These oxidation reactions are generally exothermic, raising the exhaust gas enthalpy. As a result, an earlier light-off can be achieved, improving the TWC efficiency. The amount of oxygen present in the exhaust manifolds and its mixing with unburned gas species plays a critical role in the effectiveness of the post-oxidation process. Various strategies have been proposed and studied to optimise these factors. For example, scavenging can supply the necessary oxygen by adjusting the variable valve timing (VVT), allowing fresh air to flow from the intake ports to the exhaust manifolds without contributing to in-cylinder combustion [37]. Alternatively, an air pump can inject secondary air into the exhaust system [57]. The following chapter will provide a comprehensive overview of post-oxidation in ICE. It will begin by discussing the fundamentals of post-oxidation, analysing the chemical kinetics and thermodynamics of the process. Then, a detailed examination of the reaction mechanisms involved in 3D-CFD simulations is explored. The chapter will conclude by examining various post-oxidation strategies, such as lambda-split, late fuel injection, increased secondary air injection temperature, and different exhaust manifold design concepts.

M. Pipolo, *3D-CFD Virtual Development of Post-Oxidation Strategies with Secondary Air Injection for Advanced Internal Combustion Engines*, Wissenschaftliche Reihe Fahrzeugtechnik Universität Stuttgart,
https://doi.org/10.1007/978-3-658-51343-6_3

3.1 Fundamentals of post-oxidation

Thermodynamics and chemical kinetics provide complementary perspectives on the oxidation process, which is crucial for understanding and optimising it. Chemical thermodynamics focuses on the equilibrium state, offering steady-state information about the system, such as the reaction products maximum temperature and the equilibrium composition. This analysis defines the theoretical limits of the system, including the energy potential and the extent of the reaction under ideal conditions. In contrast, chemical kinetics describes the system transient behavior, capturing the dynamics of how the system evolves toward equilibrium. Kinetics provides critical insights into the rates at which reactants (fuel and oxidiser) are consumed, intermediates are formed, and products are generated. It also quantifies the heat release rate from chemical reactions, which is fundamental in sustaining oxidation. The released heat acts as a positive feedback mechanism: it raises the temperature of the reactants, accelerating the reaction rate and driving the process forward. The interplay between thermodynamics and kinetics is of particular importance: while thermodynamics sets the ultimate limits of the reaction, kinetics determines whether and how quickly these limits can be approached. Similarly, the time required to achieve near-complete oxidation is governed by kinetic factors, especially critical in high-speed or transient systems. In the following paragraphs, the necessary information to understand the post-oxidation process from both a kinetic and thermodynamic perspective will be provided. For any additional details or further insights, the reader is encouraged to consult the literature [26, 41, 44, 64].

3.1.1 Chemical thermodynamic

Chemical reactions are always accompanied by energy changes, most commonly observed as heat absorption or heat released. These energy transformations are crucial for understanding oxidation processes, as they directly influence the temperature profile and the composition of reaction products. The heat involved in a reaction can be quantified by applying fundamental thermodynamic principles. The internal energy of a substance is a state function, meaning it depends solely on its temperature, pressure, and phase, regardless of how it reaches that state. Consequently, the change in internal energy during a physical or chemical transformation is determined only by the systems initial and final states. This energy change can manifest in different forms, primarily as heat, work, or a combination of both. Regardless of how the energy is transferred, the total change in the system energy remains constant, in accordance with the first law of thermodynamics. For practical applications, especially in combustion processes, the phenomena are often analysed through enthalpy changes ΔH rather than internal energy. Assuming that changes in kinetic and potential energy are negligible and that no work is done other than maintaining the flow, the amount of heat added to the system is equal to the change in enthalpy as the system goes from the initial to the final state ΔH (eq. 3.1):

$$Q = \Delta H \qquad \text{eq. 3.1}$$

For exothermic reactions, like combustion, the heat released Q is given by eq. 3.2:

$$Q = -\Delta H \qquad \text{eq. 3.2}$$

This principle forms the foundation of thermochemical calculations for closed thermodynamic systems. One of the most critical thermodynamic properties for the description of chemical reactions is the heat of reaction ΔH_{rxn}, which represents the total enthalpy variation that the system undergoes as the reactants transform into reaction products under standard conditions. These standard conditions are typically defined as a temperature of 298 K, a pressure of 1 atm, and the reactants and products in their most stable physical states at this temperature and pressure. The heat of reaction is connected to the heat of formation ΔH_f^0. of reactants and products. The heat of formation is the enthalpy change when one mole of a compound is synthesised from its constituent elements.The overall enthalpy change can be calculated as shown in eq. 3.3:

$$\Delta H_{rxn} = \sum \Delta H_f^0(products) - \sum \Delta H_f^0(reactants) \qquad \text{eq. 3.3}$$

This relationship highlights the energy balance driving the reaction in combustion systems. In the contex of fuel oxidation reactions, the highly negative heats of formation for stable combustion products, such as CO_2 and H_2O, reflect the significant energy release during fuel oxidation, which sustains combustion by raising temperatures and maintaining reaction rates. The heat of reaction can be either:

- Negative (exotermic): energy is released, as in combustion reactions, which sustain themselves through the heat produced.
- Positive (endotermic): energy is absorbed in specific decomposition reactions.

In combustion, the heat of the reaction is intrinsically linked to the calorific value of the fuel, which determines the energy available for practical applications. This calorific value includes the lower heating value LHV and the higher heating value HHV, essential for assessing fuel efficiency in engines and other combustion systems. The heat of reaction represents the energy released during combustion, making it a crucial factor in designing efficient systems. Fuels with higher calorific values release more energy per unit mass, enhancing performance and energy output. Furthermore, the heat of the reaction also influences post-oxidation processes that occur after the primary combustion phase. Following a simplified approach to combustion at temperatures below 1250 K, the stable products primarily include CO_2, H_2O, N_2 and O_2. In such cases, the molar quantities of these products can be determined using straightforward mass balance equations. However, in most practical combustion scenarios, the temperatures often exceed 1250 K, leading to the dissociation of stable species. These significantly endothermic dissociation reactions can reduce the flame temperature even with small levels of dissociation. The reactions that commonly occur in a $C-H-O$ system include:

$$CO+\frac{1}{2}O_2 \leftrightarrow CO_2 \qquad \text{eq. 3.4}$$

$$H_2+\frac{1}{2}O_2 \leftrightarrow H_2O \qquad \text{eq. 3.5}$$

$$H_2O+H \leftrightarrow OH \qquad \text{eq. 3.6}$$

$$H_2O+H \leftrightarrow H_2+OH \qquad \text{eq. 3.7}$$

$$O_2 \leftrightarrow 2O \qquad \text{eq. 3.8}$$

These reactions are equilibrium reactions, meaning that at any given temperature, each species involved has a specific equilibrium concentration. At high temperatures, these dissociation reactions play a crucial role in the post-oxidation processes by altering the composition of combustion products and influencing the energy balance. In low-temperature combustion, the heat of the reaction can often be calculated directly from the stoichiometric product compositions. However, including dissociation reactions adds complexity to these calculations, as both the flame temperature and the equilibrium concentrations of species become independent. Solving these requires additional equations beyond mass balances. These equations arise from equilibrium relationships, which depend on thermodynamic properties such as the Gibbs free energy G and equilibrium constants. The equilibrium state of a reacting system is determined by minimising the Gibbs free energy, given by eq. 3.9

$$G = H - TS \qquad \text{eq. 3.9}$$

where H is enthalpy, T is temperature, and S is entropy. It is significant for predicting the spontaneity of a chemical reaction:

- If $\Delta G < 0$: the reaction is spontaneous, it can proceed without additional energy input.
- If $\Delta G > 0$: the reaction is not spontaneous and requires energy input to process.

For a chemical reaction at constant temperature and pressure, eq. 3.10 expresses the condition for equilibrium.

$$(dG)_{T,P} = 0 \qquad \text{eq. 3.10}$$

This indicates that a reaction progresses toward decreasing Gibbs free energy until the chemical potentials of the reactants equal those of the products.

3.1.2 Chemical kinetics

The chemical expression of an elementary reaction is described by eq. 3.11:

$$aA + bB \leftrightarrow cC + dD \qquad \text{eq. 3.11}$$

where A and B are reactants, C and D are products, and a, b, c, and d are their stoichiometric coefficients. The rate of the forward reaction is given by eq. 3.12:

$$k_f[A]^a[B]^b \qquad \text{eq. 3.12}$$

k_f is the Arrhenius rate constant, which depends on the temperature and the nature of the reaction:

$$k_f = A_0 \exp\left(-\frac{E_a}{RT}\right) \qquad \text{eq. 3.13}$$

where:

- A_0 is the pre-exponential factor.
- E_a is the activation energy.
- R is the universal gas constant.
- T is the absolute temperature of the system.

The exponential term indicates the probability of reactant molecules overcoming the activation energy barrier during collisions. The rate of the reverse reaction is given by eq. 3.14:

$$k_b[C]^c[D]^d \qquad \text{eq. 3.14}$$

At equilibrium, the forward and reverse reaction rates are equal, giving the possibility to calculate a constant of equilibrium K, defined as:

$$K = \frac{k_f}{k_b} = \frac{[C]^c_{\text{eq}}[D]^d_{\text{eq}}}{[A]^a_{\text{eq}}[B]^b_{\text{eq}}} \qquad \text{eq. 3.15}$$

Regardless of the reaction mechanism or the number of steps involved, this equilibrium constant K is a key parameter that reflects the balance between the forward and reverse reactions. It provides an insight into the equilibrium composition of the system at a given temperature and pressure. The equilibrium constant K is directly related to the Gibbs free energy change of a reaction as by eq. 3.16:

$$\Delta G^0 = -RT \ln K \qquad \text{eq. 3.16}$$

R is the universal gas constant and T is the absolute temperature. This relationship allows the determination of equilibrium compositions of combustion products by connecting the thermodynamic properties of the system with the observed partial pressures or concentrations of species. Dissociation reactions such as those listed above are governed by their respective equilibrium constants in high-temperature combustion systems. These constants dictate how stable products like CO_2 and H_2O dissociate into intermediate species and the reverse processes during post-oxidation. Understanding the equilibrium constants is essential for modelling species redistribution in the post-oxidation phase and predicting temperatures, energy outputs and emission characteristics.

3.2 Chemical reaction mechanism

Chemical reaction mechanisms are essential for understanding the detailed pathways of chemical transformations in combustion and post-oxidation. These mechanisms include a network of elementary reactions that describe the complex interplay among reactants, intermediates and products. For example, a typical combustion process can be represented by the global stoichiometric eq. 3.17, which provides a high-level overview of the reactants and products:

$$\begin{aligned} &\mathrm{C}_a\mathrm{H}_b\mathrm{O}_g + \left(a + \frac{b}{4} - \frac{g}{2}\right)(\mathrm{O}_2 + 3.76\mathrm{N}_2) \rightarrow \\ &\rightarrow a\mathrm{CO}_2 + \frac{b}{2}\mathrm{H}_2\mathrm{O} + 3.76\left(a + \frac{b}{4} - \frac{g}{2}\right)\mathrm{N}_2 \end{aligned} \qquad \text{eq. 3.17}$$

However, the actual chemical kinetics involved in the combustion process are far more complex and do not follow such a simple or direct pathway. In reality, combustion consists of numerous intermediate steps, where various reactive species are formed, transformed, and consumed in a dynamic sequence of elementary reactions. These reactions occur at different rates and are influenced by factors such as temperature, pressure and the presence of catalytic surfaces, making the process more intricate than what the global stoichiometric equation suggests. Even for simple systems like the combustion of hydrogen and oxygen, the overall stoichiometric equation can be written as eq. 3.18, representing only the net result of a series of intricate chemical reactions above-mentioned:

$$H_2 + 0.5O_2 \rightarrow H_2O \qquad \text{eq. 3.18}$$

The reaction does not involve direct interaction between H_2 and O_2. The simultaneous breaking of $H-H$ and $O-O$ bonds are highly improbable during a single molecular collision. Instead, the reaction begins with generating highly reactive intermediates known as radicals. The initiation of the reaction typically involves the dissociation of molecules with the help of a third body M, which acts as an energy carrier, as shown in eq. 3.19 and eq. 3.20:

$$H_2 + M \rightarrow H + H + M \qquad \text{eq. 3.19}$$

$$O_2 + M \rightarrow O + O + M \qquad \text{eq. 3.20}$$

Radicals such as H and OH are then involved in a chain of reactions that sustain and propagate the combustion process. For instance, H radicals react with O_2 to form HO_2 and other intermediates, which in turn participate in further reactions, building up a network of reactive radicals that accelerate the reaction rate. This sequence of elementary reactions constitutes what is known as a reaction mechanism. Depending on the level of detail required, reaction mechanisms can range from semi-global models with a few steps to highly detailed mechanisms involving thousands of reactions. For instance, hydrogen combustion can be described by 9 species and 21 elementary reactions, while methane combustion requires 53 species and 400 reactions (e.g., the GRI 3.0

mechanism [29]). For more complex fuels like isooctane, mechanisms may involve up to 860 species and over 3600 reactions. These detailed mechanisms are essential for accurately modelling the chemical kinetics of combustion, but they also demand substantial computational resources. Elementary reactions in combustion can be categorised into four main types, each playing a crucial role in the reaction mechanism:

- Chain initiation: these reactions generate radicals to start the combustion process.
- Chain propagation: these reactions maintain the radical population without increasing or decreasing it.
- Chain branching: these reactions increase the number of radicals, accelerating the reaction.
- Chain termination: these reactions reduce the number of radicals, slowing the reaction.

In combustion scenarios characterised by a fuel-rich mixture (λ<1), incomplete oxidation occurs due to insufficient oxygen to oxidise the fuel entirely. Under such conditions, the fuel still interacts with the flame front and initiates the chemical chain reactions previously described, such as chain initiation, propagation, and branching. However, these reactions cannot proceed to completion due to the oxygen deficiency. As a result, the combustion process produces significant quantities of intermediate species such as CO and H_2 instead of complete combustion products like CO_2 and H_2O [26, 64]. These intermediates form because the chemical pathways favour partial oxidation processes in a fuel-rich environment, where the reactions are primarily controlled by the limited oxygen available. For example, CO is often produced by the incomplete breakdown of hydrocarbons or the partial oxidation of carbon, which is thermodynamically favoured when the available oxygen quantity is low. Similarly, H_2 emerges due to the dehydrogenation of hydrocarbons and other chain propagation steps involving radicals like OH and H [41]. A reasonable approximation for fuel-rich combustion systems is to assume that the unburned or partially oxidised fuel is predominantly converted into CO and $H2$. This assumption greatly simplifies post-oxidation modelling in computational simulations, as it reduces the complexity of the chemical reaction mechanism while still representing the dominant chemical pathways governing incomplete oxidation. Indeed, while actual fuel combustion processes may involve hundreds or even thousands of elementary reactions and intermediate species, focusing on CO and H_2 as primary intermediates offers an effective trade-off between computational efficiency and modelling accuracy [14]. For example, studies such as those conducted by Tromellini [63] demonstrated that including reduced chemical reaction mechanisms, like the $Polimi\_1412\_H_2/CO$ mechanism [14], can provide accurate results for post-oxidation simulations while keeping computational costs manageable. This mechanism, which includes 14 species and 33 reactions, was explicitly optimised for tracking the oxidation of CO and H_2, making it a suitable choice for scenarios involving incomplete combustion. Therefore, while simplified assumptions such as considering CO and H_2 as the dominant intermediates may overlook certain minor species, they benefit post-oxidation models in 3D-CFD simulations. When studying post-oxidation strategies that involve secondary fuel injection, the requirements for the reaction mechanism become significantly more complex. Secondary fuel injection, often performed shortly before the exhaust valve opens, introduces unburned fuel directly into the post-oxidation environment. This necessitates a detailed reaction mechanism to accurately model the conversion of unburned fuel into fully oxidised products. In these scenarios, the chemical pathway is more intricate, requiring the inclusion of additional species and reactions that describe the oxidation of hydrocarbons THC and other complex intermediates. The reaction mechanism previously employed in simulations, such as the $Polimi\_1412\_H_2/CO$, must be updated to include a more detailed model due to the complexity of such phenomena. For instance, the $Polimi\_1412\_detailed_mechanism$ [51] consists of 156 species and 3465 reactions, offering a more comprehensive representation of the post-oxidation process. While this level of detail ensures greater accuracy, it also significantly increases computational demand, requiring careful consideration of simulation resources. When employing secondary injection, the injected fuel can be assumed to be unburned at the moment of injection. This necessitates using a surrogate fuel to represent the chemical composition of the actual fuel in simulations. Directly modelling the accurate fuel composition is impractical due to the large number of reactions involved, which would lead to excessive computational times. To balance computational feasibility with chemical accuracy, a surrogate fuel is defined based on representative components and their proportions. In this study, the surrogate fuel composition is derived from the work of Cupo [15], and includes four primary components: toluene, n-heptane, ethanol, and isooctane. the specific amounts of these components in the fuel surrogate are reported in 3.1, providing a reasonable approximation of the actual fuel chemical behaviour.

Table 3.1: Surrogate composition for post-injection simulations

Component	Isooctane	Toluene	N-heptane	Ethanol
Vol. percentage	40%	36%	13%	11%

3.3 Overview of post-oxidation strategies

Post-oxidation refers to the oxidation of incomplete combustion products, CO, H_2 and HC, with oxygen introduced into the exhaust manifold. This process occurs downstream of the combustion chamber, typically during the exhaust stroke, and can significantly influence emissions control and thermal efficiency in internal combustion engines. Post-oxidation strategies can be categorised based on how oxygen and fuel are supplied to the exhaust system to facilitate the oxidation of rich combustion products. The primary methods include:

1. Oxygen supply strategies:
   - Scavenging: fresh air naturally enters the exhaust manifold during valve overlap, providing the oxygen required for post-oxidation. This process leverages the in-cylinder flow dynamics and valve timing to enhance oxygen availability. Scavenging is a passive strategy but depends heavily on the engine operating conditions and the timing of intake and exhaust valves [63]
   - Secondary air injection: in this active strategy, additional air is injected into the exhaust manifold to supply oxygen for post-oxidation. Secondary air injection systems can be precisely controlled to optimise oxygen delivery, ensuring efficient CO, H_2, and hydrocarbon oxidation. This strategy is particularly effective under transient operating conditions where scavenging alone may be insufficient.

2. Fuel supply strategies:

    - Lambda split: in this approach, the air-fuel ratio is deliberately different among cylinders or during specific engine cycles. Some cylinders operate under rich conditions to produce the necessary combustion products, while others operate under lean or stoichiometric conditions to enhance oxygen availability. The mixing of exhaust gases from these cylinders promotes post-oxidation.
    - Rich combustion: rich in-cylinder combustion naturally produces significant CO, H_2, and unburned hydrocarbons that flow into the exhaust system. While this approach does not involve additional fuel supply, it relies on effectively mixing fresh air (from scavenging or secondary injection) with these combustion products.
    - Secondary fuel injection: secondary fuel injection involves injecting additional fuel directly into the cylinder, with the timing of this injection playing a crucial role in determining the reactivity and distribution of the fuel. When the injection occurs shortly before the ignition point, the secondary fuel is introduced just before the combustion event. Alternatively, the injection can occur before the exhaust valves open, allowing the unburned fuel to be discharged into the exhaust manifold during the exhaust stroke. Another possible strategy involves injecting the fuel just before the exhaust valve closes, near the end of the exhaust stroke. In this case, the injected fuel remains within the cylinder and becomes available for the subsequent combustion cycle.

3. Other potential strategies:

    - Manifold design and flow optimization: the geometric design of the exhaust manifold plays a significant role in post-oxidation efficiency. Optimised manifold designs can promote better mixing of exhaust gases and fresh air, increase residence time, and maintain high temperatures, all of which are critical for post-oxidation.

The extent and efficiency of post-oxidation are predominantly influenced by four key factors: mixing conditions, the availability of oxygen and the presence of rich combustion products, the temperature and residence time. Efficient mixing between oxygen (supplied by fresh air) and rich combustion products is vital for post-oxidation. In the exhaust manifold, the fresh air and exhaust gases exist as separate clouds in a highly turbulent environment. The interaction between these clouds at their interface creates the reaction zones where oxidation takes place. The turbulence, especially during the exhaust stroke, improves mixing by increasing the surface area available for reactions. However, uneven mixing can create oxygen-rich and oxygen-lean regions, which reduces overall oxidation efficiency. Achieving a balance between turbulence and reactant interaction is therefore essential to maximize post-oxidation performance. The presence of oxygen is an indispensable prerequisite for post-oxidation reactions. It can be introduced into the exhaust manifold through scavenging, where fresh air enters during valve overlap, or via secondary air injection, where additional air is deliberately supplied. The oxygen available determines the extent to which incomplete combustion products are further oxidized into CO_2 and H_2O. Without sufficient oxygen, post-oxidation remains incomplete, resulting in elevated emissions of unburned species. Similarly, the availability of rich combustion products is a critical factor, as post-oxidation requires the presence of incomplete combustion products in the exhaust gases. These products, primarily formed under fuel-rich conditions due to insufficient in-cylinder oxygen, are partially oxidized species capable of undergoing further oxidation when exposed to oxygen in the exhaust manifold. CO and H_2 are particularly significant because their oxidation reactions occur rapidly at sufficiently high temperatures. In contrast, hydrocarbons often demand more time and higher activation energies for oxidation due to their complex molecular structures. Temperature is another critical parameter, with oxidation reactions for CO and H_2 requiring a minimum temperature of approximately 1000 K to overcome activation energy barriers. In practical engine cycles, the gas temperature in the exhaust manifold fluctuates significantly. During the exhaust stroke, the hot combustion products entering the manifold can typically reach these critical temperatures, enabling post-oxidation. However, as the gases expand and mix with colder air, the temperature may drop below the required threshold, inhibiting further oxidation. Residence time, the duration that reactants remain in high-temperature zones, is equally crucial, as post-oxidation reactions are time-dependent. Insufficient residence time can inhibit complete oxidation, limiting the efficiency of the process.

4 Methodology and Simulation Tools

The development of sustainable ICEs faces many challenges, particularly in understanding and optimising complex phenomena such as post-oxidation. Achieving a balance between performance, emission reduction and efficiency requires advanced experimental techniques and numerical tools that allow a deeper investigation of the combustion and exhaust processes. In the context of this work, a four-cylinder, turbocharged, gasoline direct injection spark ignition engine has been investigated. Figure 4.1 illustrates a work process that relies on accurate measurements taken at the test bench by Chiba University, verified through 3D-CFD simulations at the University of Stuttgart. The joint efforts of these two institutions have been key in providing valuable insights into the research topic, integrating knowledge from both experimental and computational areas. Measuring and visualising post-oxidation at the test bench is a highly challenging task due to the inability to directly measure the heat release rate from oxidation reactions. Therefore, 3D-CFD simulations are essential in interpreting experimental results and optimising the system considering new engine configurations or different operating strategies, ensuring cost and time reduction within the development process. Two types of numerical approaches have been implemented for the investigations: a full engine simulation and a detailed model of the exhaust manifold with the kinetic reaction mechanism. The 3D-CFD full engine simulation accomplished in the 3D-CFD tool *QuickSim* [13] provides a better understanding of the overall engine behaviour and the phenomena affecting the oxidation, i.e. inhomogeneities in the mixture. This approach ensures the implementation of ambient boundary conditions, thereby avoiding the specification of any three-dimensional flow, which is particularly important in 3D-CFD simulations. The choice of boundary conditions significantly influences the accuracy of the results, as traditional approaches relying on measured or 1D-CFD-calculated boundary conditions often assume a uniform distribution of flow variables. However, in reality, the intake and exhaust regions are characterized by complex three-dimensional flow patterns, including mixing, heat exchange, and pressure waves. By applying ambient boundary conditions, this approach minimizes assumptions and provides a

M. Pipolo, *3D-CFD Virtual Development of Post-Oxidation Strategies with Secondary Air Injection for Advanced Internal Combustion Engines*, Wissenschaftliche Reihe Fahrzeugtechnik Universität Stuttgart,
https://doi.org/10.1007/978-3-658-51343-6_4

more consistent and physically representative initialization of the flow field, ultimately improving the reliability of post-oxidation simulations.

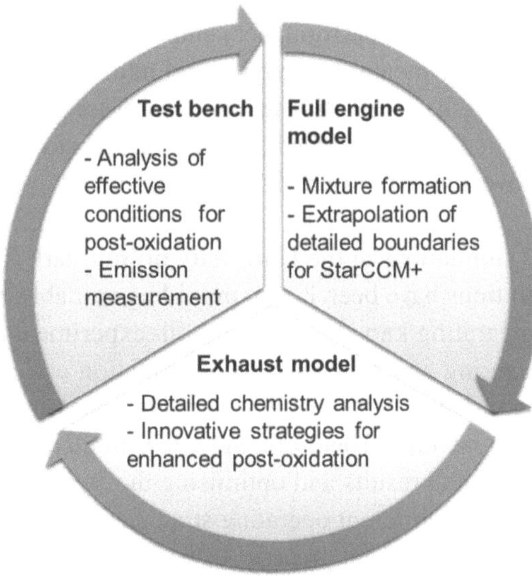

Figure 4.1: Investigations working flow.

Several engine operating points have been run and validated against experimental data. Consequently, reliable output (exhaust gas composition and distribution) can be extrapolated from the complete model and adopted as input boundary conditions for the detailed exhaust simulation performed in *StarCCM+* [58]. Using the input species concentration and the kinetic reaction mechanism, it is possible to analyse the main parameters influencing the post-oxidation, such as temperature, mixing conditions, availability of rich combustion product and oxygen. Furthermore, it is possible to collect the heat release rate resulting from oxidation reactions of the rich combustion products, which is a highly challenging task at the test bench. Analysing post-oxidation through a dedicated model of the exhaust manifold is efficient in terms of time, as the 3D-CFD tool *QuickSim* already investigates the overall engine behaviour and combustion. The first section of this chapter describes the experimental setup and the role of the test bench in data acquisition. Subsequently, the 3D-CFD tool *QuickSim* is introduced highlighting its capabilities and contributions to the study of post-oxidation. Finally, a detailed analysis of post-oxidation phenomena using *StarCCM+* is presented, focusing on the exhaust manifold and kinetic reaction mechanisms.

4.1 Experimental setup

The investigation was performed on a four-cylinder turbocharged gasoline direct-injection spark-ignition engine, whose specifications are reported in Table 4.1. The engine was coupled with a low-inertia Horiba dynamometer to simulate real-world driving conditions. The test bench was equipped with a range of advanced measurement instruments and sensors, enabling the precise acquisition of data necessary to achieve the objectives of this study. The configuration used is shown in Figure 4, and the engine specifications are provided in Table 1. Engine speed and load were controlled using a dynamometer controller. At the same time, in-cylinder pressure was measured with a piezoresistive pressure transducer (Kistler 6041B), with the resulting signal converted to voltage through a charge amplifier integrated into the KiBox system. Exhaust pressure was monitored using exo-resistive pressure transducers (Kistler K4005BA5F and Kistler 4624A), which recorded data at 0.1 crank angle intervals.

Table 4.1: Engine specifications.

No. of cylinders	4, Inline
Volume capacity, l	1.6
Bore x stroke, mm	79.7 x 81.1
Compression ratio	10.5
Max power, kW/rpm	140/5600
Max torque, Nm/rpm	240/1600-5200
Valve train	DOHC 16 valves
Fuel injection	Direct injection

All other parameters were recorded using a low-speed data logger, with measurements taken at 0.5 seconds intervals over one minute and the resulting values averaged. Key temperatures relevant to engine performance were monitored using K-type thermocouples, including temperatures of intake air, exhaust gas, coolant and oil. Exhaust gas composition was analysed using the Mexa 1600 exhaust gas analyser, with lambda values derived directly from calculations performed within the analyser. An air-fuel (A/F) ratio sensor was installed upstream the catalyst to provide feedback control for maintaining the theoretical air-fuel ratio via the ECU. The locations of the measurement devices are illustrated in Figure 4.2. Exhaust pressure was measured at ports one and two (belonging, respectively, to cylinder one and two). In contrast, exhaust gas temperature and composition were sampled at multiple points: within each exhaust port before the turbocharger, downstream of the turbocharger, and after the catalyst.

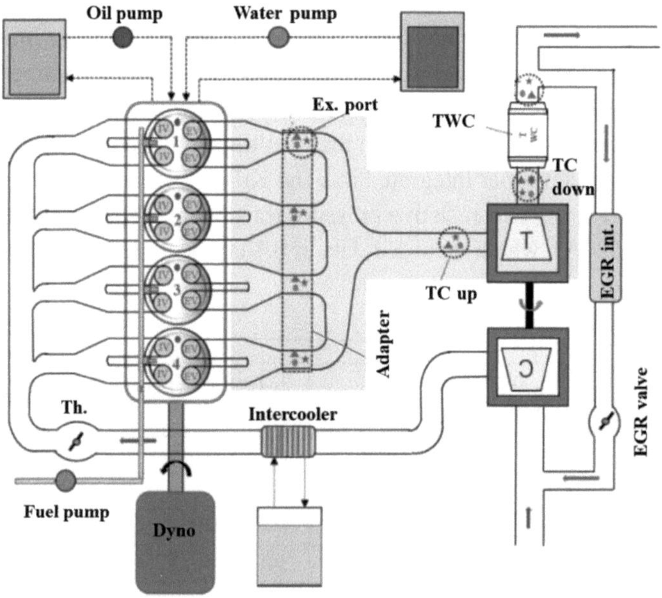

Figure 4.2: Measurements position.

The secondary air injection system was installed at the engine exhaust port, as shown in Figure 4.3. This system included a compressor, a pressure regulator, a volumetric flow meter and a volumetric flow control regulator, ensuring precise air delivery near the exhaust valves. The SAI tube diameter was set at 6 mm based on considerations of the required mass flow rate, acceptable pressure drop, and insights from prior studies. Each SAI tube connected to the exhaust ports was equipped with a one-way valve to prevent the backflow of combustion gases into the tube during conditions where exhaust manifold pressure exceeded the SAI tube pressure. For instance, at an intake air volume of 10 g/s, an SAI rate of 10% corresponds to an injection of 1 g/s of secondary air.

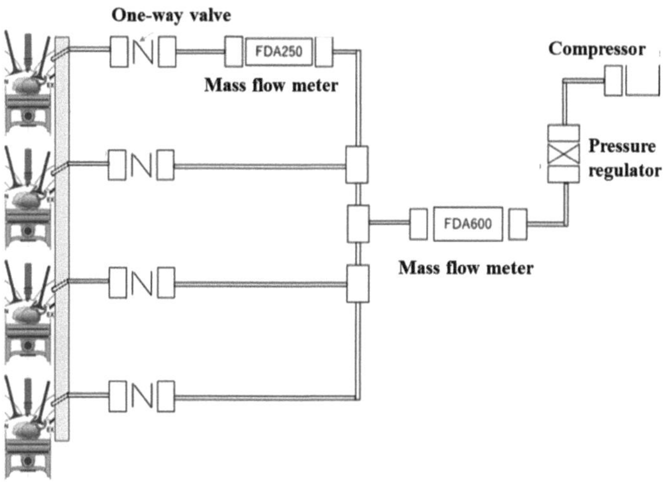

Figure 4.3: Secondary air injection system.

4.2 The 3D-CFD fast-response tool QuickSim

The numerical investigations presented in this work were conducted using the 3D-CFD simulation tool *QuickSim* [13], a fast-response 3D-CFD tool developed and continuously refined over the past two decades by the Forschungsinstitut für

Kraftfahrwesen und Fahrzeugmotoren Stuttgart (FKFS) in collaboration with the Institut für Fahrzeugtechnik Stuttgart (IFS) at the University of Stuttgart. It employs the commercial software STAR-CD as a solver and incorporates ICE-optimized numerical models to achieve high simulation accuracy while minimising computational requirements. One of the distinguishing features of the 3D-CFD tool *QuickSim* is its ability to utilise coarser computational meshes compared to traditional 3D-CFD approaches. This is made possible through ICE-specific phenomenological models that are custom-made and improved for such applications. For instance, a complete operating cycle of a single cylinder engine model can be simulated in approximately two hours using a 12-core CPU. This efficiency enables the simulation of multiple successive operating cycles, extending to transient conditions, within a practical time frame. To understand the time-saving advantages offered by *QuickSim*, Figure 4.4 summarises the typical time scales ranging from real-time electronic control unit processes to various CFD methodologies.

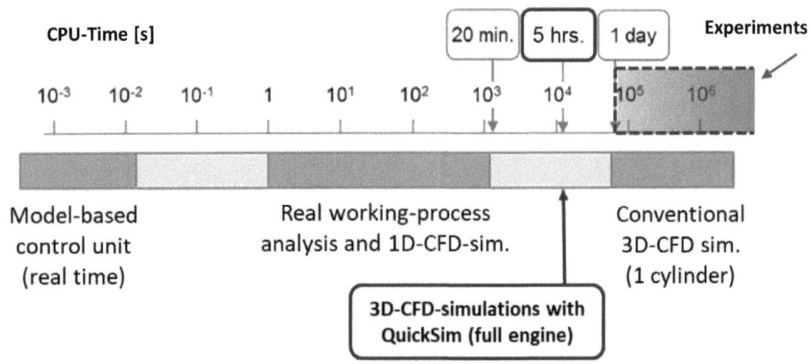

Figure 4.4: Typical time scales for simulations and real time controls.

The methodology of the 3D-CFD tool *QuickSim* provides several advantages. It allows the entire engine to be included in the simulation domain, surrounded by the intake and exhaust systems. The simulation begins with an initial estimation of flow fields and thermodynamic conditions, progressing through a transient phase until stable operating conditions are achieved. Typically, this transient phase concludes after 5–10 cycles, depending on the engine operating point and specific model configurations. At this stage, the influence of user-defined initial conditions is minimised, enabling a realistic representation of engine behaviour. The fast-response 3D-CFD tool employs previously calculated look-up tables for combustion properties to further enhance computational efficiency. These properties, including laminar flame speed and ignition delay time, are evaluated in a preliminary phase using detailed reaction mechanisms in Cantera [28]. The resulting data, such as caloric properties across varying temperatures, pressures, and mixture compositions, eliminate the need for real-time chemistry calculations during simulations. This approach is particularly advantageous for reducing computational load while accurately predicting phenomena like auto-ignition. The main numerical models implemented in *QuickSim* for this study are summarised in Table 4.2, with detailed descriptions in [13].

Table 4.2: Overview of the 3D-CFD tool *QuickSim* numerical models.

Simulation methodology	Multi-phase RANS
Turbulence modelling	k-ε
Multi-phase flow	Euler / Lagrange
Combustion - Flame propagation	Two-zone Weller-model adapted to GDI-combustion for spark ignition flame propagation and self-ignition model
Solver	Coupled flow and energy with 2nd order implicit method
Time modelling	Stationary

In the 3D-CFD tool *QuickSim*, the working fluid is represented through a system of scalars rather than individual chemical species. These scalars are designed to capture the primary groups of interest in engine operation, including fresh charge, burned gas and EGR. This allows for a simplified and effective representation of the combustion process. This approach significantly reduces computational effort compared to explicitly modelling each chemical species while maintaining sufficient engine simulation accuracy. For clarity, figure 4.5 illustrates the composition of a typical 3D-CFD cell during combustion, where the flame front separates the burned and unburned zones. Table 4.3 summarises the definitions of these scalars, while additional details are available in the literature [13].

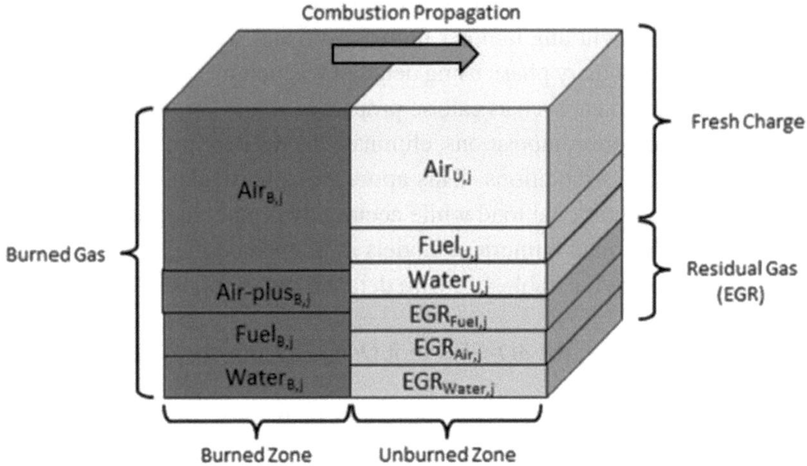

Figure 4.5: Scalars composing a computational cell in *QuickSim*.

Table 4.3: Description of the base scalars defined in *QuickSim*.

Scalar	**Description**
$w_{Air_{U,j}}$	Mass fraction of fresh air
$w_{Fuel_{U,j}}$	Mass fraction of fresh vaporized fuel
$w_{EGR_{Air,j}}$	Mass fraction of air that has previously produced EGR (burned gas of the previous operating cycle)
$w_{EGR_{Fuel,j}}$	Mass fraction of vaporized fuel that has previously produced EGR (burned gas of the previous cycle
$w_{Air_{B,j}}$	Mass fraction of air that has previously produced burned gas
$w_{Fuel_{B,j}}$	Mass fraction of vaporized fuel that has previously produced burned gas
$w_{Air_{B,plus}}$	Mass fraction of air present in the burned gas

To distinguish between exhaust gases in front of the flame front and those generated during combustion, separate scalars for burned gas and EGR are implemented. This differentiation enables the accurate representation of the local distribution of these components and the flame front dynamics. Furthermore, the introduction of parameters such as $\lambda burned$ and λEGR allows the characterisation of the air-fuel ratio (rich, lean, or stoichiometric) concerning the properties of the combustion products. This capability is significant for studying post-oxidation phenomena, where the interaction between oxygen-rich exhaust gases and unburned hydrocarbons can lead to post-oxidation events. Including a scalar called $air - burn - plus$ further enhances the ability of the model to simulate post-oxidation. This scalar accounts for fresh air in burned gases with a local air-fuel ratio exceeding stoichiometry, facilitating the analysis of oxygen availability for secondary oxidation reactions.

4.2.1 Complete engine with 0D turbocharger model

The simulation model utilised in *QuickSim* for this investigation is about 1 million cells and expands the computational domain from the air filter to the catalyst, incorporating a 0D turbocharger model (as illustrated in Figure 4.6). On the intake side, the system includes an air filter with a specified porosity value to replicate the pressure drop observed in real conditions. Following the air filter, the airflow passes through the compressor and the intercooler, where it is cooled upon exiting the compressor. The intake system also features a throttle valve, which, in principle, could be adjusted to simulate partial load conditions. However, for the purposes of this study, the throttle was considered fully open to represent full-load operation. Within the intake manifold, multiple sensors are installed in positions identical to those on the test bench, ensuring accurate comparison between simulation and experimental measurements. The engine consists of four cylinders, offset by 180 degrees, with a firing order of 1-3-4-2. On the exhaust side, secondary air injectors are located 50 mm downstream of the exhaust valves. These injectors, along with other components of the exhaust system, are depicted in detail in the corresponding detailed exhaust representation. Similar to the intake system, the exhaust manifold is equipped with sensors precisely positioned to allow a direct comparison with test bench data. Finally, the exhaust system includes the turbine and the catalyst.

The catalyst is modelled with a specific porosity to accurately replicate the pressure drop across its surface. The turbocharger model used in this study provides an efficient and robust approach for integrating turbocharging effects into the *QuickSim* simulation environment. Based on performance maps of the compressor and turbine, the model captures the relationship between key parameters such as mass flow, pressure ratio and efficiency under varying operating conditions. These maps are loaded into memory at the start of the simulation, allowing for rapid access during runtime. The model is well integrated with the 3D computational domain of the engine simulation, enabling continuous data exchange at the internal boundary conditions (i.e. turbine inlet, turbine outlet in figure 4.6). Flow information such as pressure, temperature, and species concentration is collected from the 3D domain and converted into reduced conditions using predefined equations. This facilitates the calculation of the turbocharger instantaneous performance, including power and torque output. The rotor dynamics are modeled by calculating the net torque on the turbocharger shaft, which is derived from the contributions of the turbine and compressor, as well as frictional losses. The resulting torque determines the shaft speed, which is then used to update the performance maps in real time. This dynamic interaction ensures accurate representation of the coupled behavior between the engine and the turbocharger. The 0D model simplifies the process by treating the turbocharger as a black box, allowing for the simulation of complex phenomena, such as reverse flow and compressor surge, with high efficiency while maintaining a satisfactory level of accuracy. The integration of the turbocharger into the simulation domain eliminates the need for external simulations to define boundary conditions. This simplification reduces the complexity of the model, as ambient pressure and temperature at the domain boundaries are sufficient, avoiding the inaccuracies associated with predefined boundary conditions. For any additional detail, the reader can refer to [34].

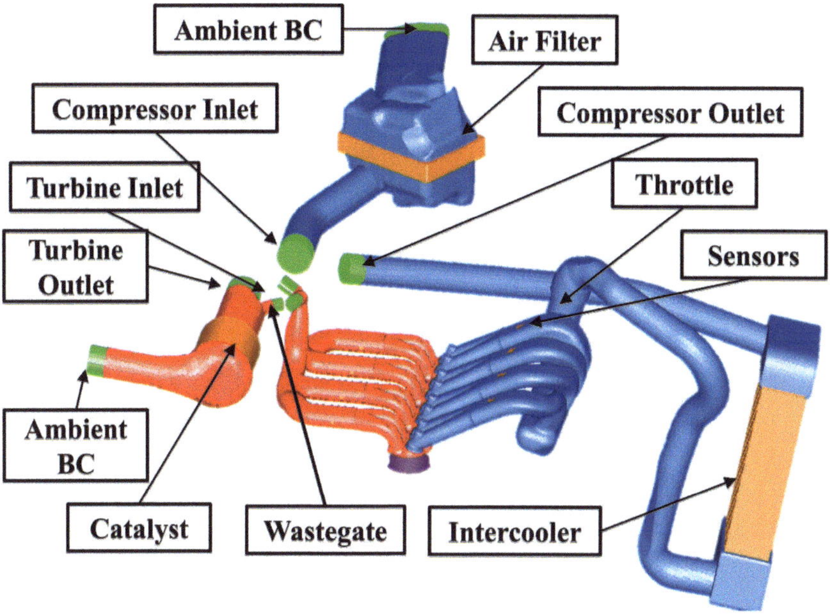

Figure 4.6: Full engine model developed in *QuickSim*.

The most important aspects to consider for the post-oxidation are time, mixing conditions, high temperatures, and availability of oxygen-rich combustion products. In order to improve the accuracy of the mixing condition prediction, a final development step in the project has been the integration of a 3D turbocharger model to enable a complete and realistic replication of the turbine. This is particularly critical because post-oxidation processes can also occur downstream of the turbine. Unlike the 0D turbocharger model, which imposes uniform boundary conditions at the turbine inlet, the 3D model accounts for the non-uniformity of the flow field, ensuring a more accurate interaction between the exhaust gases and the turbine. This refinement not only improves the representation of mixing processes downstream but also influences the turbine upstream, as it eliminates the artificial assumption of uniform flow near the internal boundary condition. This topic will be discussed in greater detail in Chapter 7.

4.3 Detailed simulation using the 3D-CFD software StarCCM+

Using the commercial 3D-CFD tool *StarCCM+*, the second type of simulation focuses on modelling the exhaust manifold from the exhaust valves up to the turbine inlet, including the secondary air injectors and the 10 cm adaptor between the cylinder head and the four-in-one exhaust manifold, as shown in figure 4.7. This model is characterised by a finer mesh with 400000 cells and a chemical reaction mechanism integration.

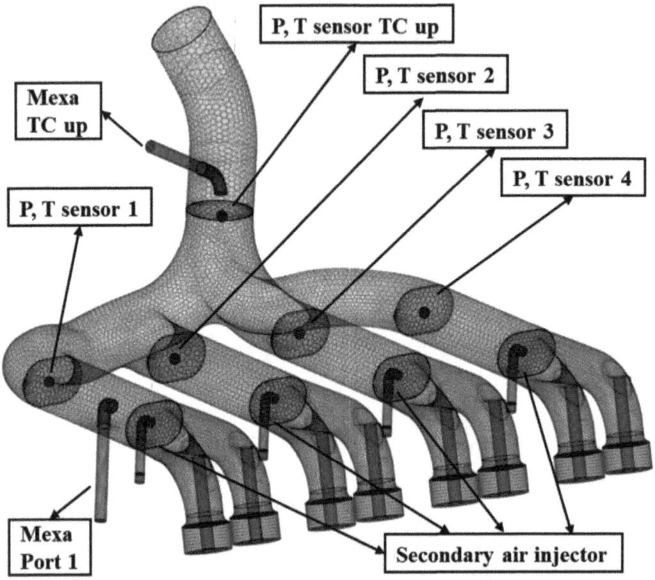

Figure 4.7: *StarCCM+* exhaust manifold model and sensors position.

To accurately replicate the test bench methodology in the simulation, it was crucial to reproduce the sampling of the species with a constant mass flow rate similar to that of the Mexa 1600 system. The model offers a detailed account of the physical and chemical processes within the exhaust manifold throughout the engine cycle and highlights the variables that have the most significant impact on post-oxidation. The advantage of this approach is that the model behaves realistically, considering all the complex phenomena examined in the full engine simulations. Table 4.4 reports an overview of the models implemented in *StarCCM+*.

Table 4.4: *StarCCM+* implemented model.

Flow	Turbulent – Unsteady RANS equations
Turbulence	k-ε
Wall treatment	Two-layer all y^+
Fluid	Multicomponent gas
Equation of state	Ideal gas

The choice of the reaction mechanism depends on the definition of the species and reactions that are needed during post-oxidation. The simulation setup utilises a reaction mechanism consisting of three main components, CO, H_2, and O_2, as shown previously in chapter 3.2. The boundary conditions given to *StarCCM+* are evaluated in the 3D-CFD tool *QuickSim* at the exhaust valve, extrapolating for each cell the mass of rich combustion products, lambda, pressure and temperature. Hence, considering the total mass flow through the valves, the concentrations are evaluated, and the detailed boundary conditions are set, as shown in Figure 4.8.

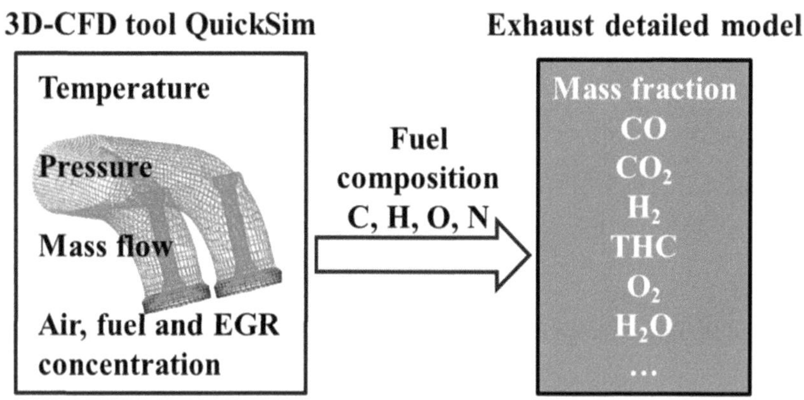

Figure 4.8: Detailed boundary extrapolation in *QuickSim*.

Temperature measurements are crucial to ensure precise post-oxidation, and comparing the temperature between the experiment and the simulation is essential. One approach is using a thermocouple model in the simulation to improve the accuracy of the 3D-CFD simulation. The implemented solution involves a numerical model that incorporates thermodynamic data obtained from simulations and information about the dimensions and materials of the utilised thermocouple [30]. This approach allows for considering the actual effects of the thermocouple, such as thermal inertia, conductive heat exchange with the surrounding walls, and convective heat exchange with the gas. Figure 4.9 illustrates the application of this model on the thermocouple 4 (refer to figure 4.7). The graph displays the measurement obtained from a standard sensor represented by a continuous line. Additionally, it depicts the application of the virtual thermocouple model to the same sensor, represented by a dashed line. Using the virtual thermocouple leads to a more stable temperature signal, which is more similar to the temperature signal of the measurements at the test bench.

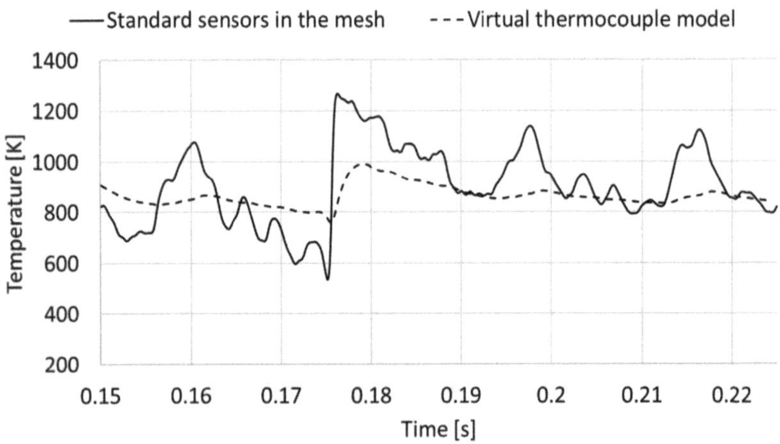

Figure 4.9: Temperature with a standard sensor and a virtual thermocouple.

4.4 Validation of the numerical approaches

The 3D-CFD simulation methodology was successfully validated by applying it to engine operating points previously tested at the test bench, listed in table 4.5. The amount of secondary air is chosen so that the lambda before the TWC is around 1. The engine points were validated through 3D-CFD simulations, and different strategies for post-oxidation were investigated. The validation process involved the full engine simulation model and the detailed model of the exhaust manifold. All the experimental information on sensors and pollutants was compared with the simulation data, confirming the methodology accuracy and reliability.

Table 4.5: Engine operating points run within the 3D-CFD environment.

RPM	Load, Nm	Air cons, kg/h	IMEP, bar	SAI, kg/h
1600	160	96.73	12.8	-
1600	160	108.6	13.4	18.4
2000	140	113.9	11.5	-
2000	140	130.3	11.5	20.8
2000	160	128.9	13.1	-
2000	160	141.6	13.2	21.2
1600	150	143.6	12.2	18.7

4.4.1 3D-CFD full engine simulations

The primary data obtained from the simulation and test bench are presented in table 4.6. Minor discrepancies between the results can be attributed to the differences in methodologies between the simulations and experimental measurements. Specifically, the air consumption values in the simulation and test bench exhibit slight differences because the 3D-CFD method accounts for the mass flow at the intake system. In contrast, the test bench measures the air consumption based on stoichiometric ratio, fuel consumption, and the global lambda at the TWC, the latter being affected by the oxygen provided by the SAI system. Hence, the oxygen supplied by SAI and its consumption due to post-oxidation plays a vital role in determining the lambda at the TWC, which can lead to an overestimation of the engine air consumption at the test bench.

Table 4.6: Engine data at 1600 rpm 160 Nm and 2000 rpm 2000 140 Nm

Variable	1600 rpm		2000 rpm	
	Test bench	Sim.	Test bench	Sim.
IMEP,bar	13.5	13.6	11.6	11.8
Air cons., kg/h	111.5	103.6	129.4	120.7
Fuel cons., kg/h	7.6	7.6	8.88	8.88
P_{max}, bar	47.4	46.8	36	35.4
MFB50, °CA a.TDC	25.8	25.7	30.5	30.2

Figure 4.10 and figure 4.10 illustrate the in-cylinder pressure signal for the 1600 rpm 160 Nm and 2000 rpm 140 Nm engine operating points, respectively. Both engine operating points are characterised by a late IP (0 deg, FTDC) to ensure that the in-cylinder gases have a sufficiently high temperature when the exhaust valves open: the higher the temperatures, the better the post-oxidation.

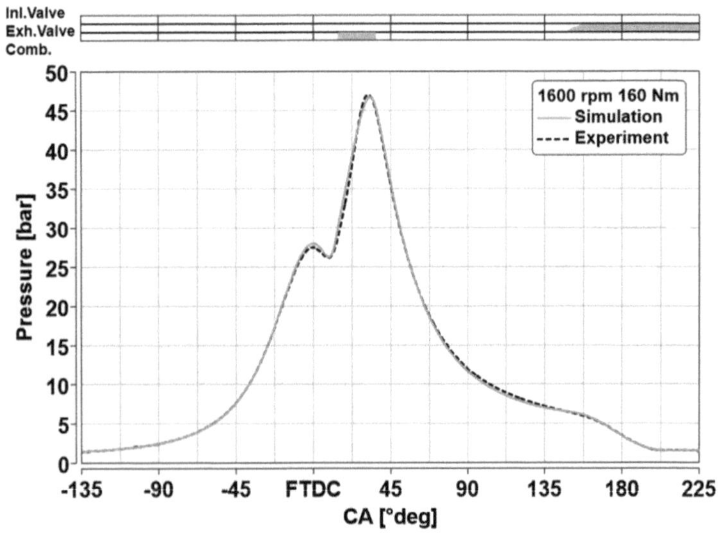

Figure 4.10: In-cylinder pressure at 1600 rpm 160 Nm.

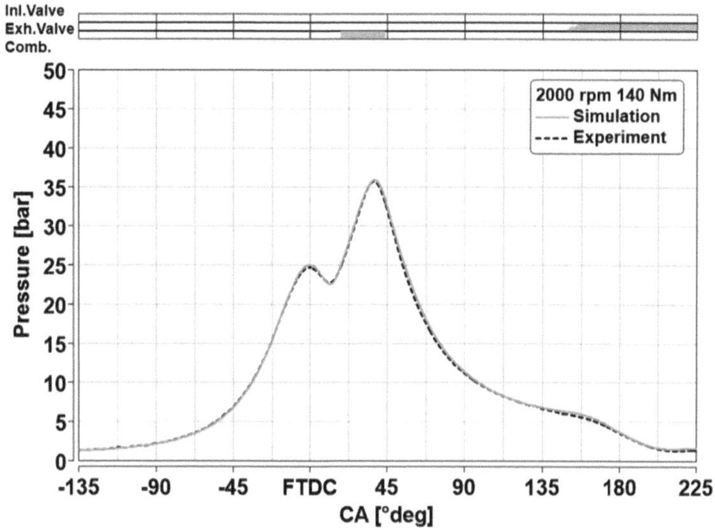

Figure 4.11: In-cylinder pressure at 2000 rpm 140 Nm.

The measured lambda on the test bench considers the O_2 level in the exhaust gases, which can be influenced by the inhomogeneity of the mixture. As a result, a locally rich mixture can mislead the lambda measurement and result in slightly different air consumption values between the simulation and test bench. On the other hand, when the engine is operating with SAI, more fuel is injected, resulting in a richer lambda (considering the same air entering the cylinder among the SAI and no-SAI cases). As a result, the differences between the test bench and simulation data can be attributed to measurement methodology. Equation eq. 4.1 is used to calculate the air consumption experimentally, but this method needs to accurately represent the actual air engine consumption since the lambda measurement is made at the TWC. The air coming from the secondary air injectors and its reduction due to post-oxidation are considered at the TWC, which can lead to overestimating the engine air consumption on the test bench. Moreover, the overestimation of the engine air consumption on the test bench highlights the importance of accurately measuring lambda values and considering the impact of local mixture inhomogeneity in engine simulations. Despite these differences, simulation and test bench data are valuable in providing insights into the engine performance and developing strategies to improve post-oxidation. The lambda value measured on the test bench is based on the O_2 level in the exhaust gases, which can be affected by the inhomogeneity of the air-fuel mixture. This means that a locally rich mixture can mislead the lambda measurement, leading to slight discrepancies in air consumption values between simulation and test bench data. During engine operation with SAI, additional fuel is injected to create a richer lambda in the cylinder, producing combustion products that can be further utilized in the exhaust for post-oxidation. Assuming that the same amount of air enters the cylinder in both SAI and non-SAI cases, the differences in the air consumption between experiment and simulation can primarily be attributed to the measurement methodology. On the test bench, air consumption is calculated using equation eq. 4.1 (kg/h), which relies on the lambda measurement after the three-way catalyst (TWC).

$$\dot{m}_a = \lambda \cdot \alpha_{st} \cdot \dot{m}_f \qquad \text{eq. 4.1}$$

The air consumption is evaluated as a function of lambda, stoichiometric ratio, and fuel consumption. The α_{st} used for the calculation is 14.3. However, this method does not fully represent the actual air consumed by the engine for two reasons:

1. It can not distinguish between the air introduced by secondary air injectors and the air consumed directly by the engine.
2. The measurement is based on the oxygen content in the exhaust gases, which is also influenced by the post-oxidation rate.

Higher post-oxidation increases the amount of oxygen consumed, leading to an incorrect evaluation of the engine air consumption at the test bench. Consequently, accurately measuring lambda values and accounting for local mixture inhomogeneity are crucial in understanding the differences between simulation and test bench data. Indeed, in the simulation, only the air entering the cylinder through the valves is considered, making it a more accurate representation of the engine air consumption.

4.4.2 3D-CFD exhaust manifold simulations

Figure 4.12 illustrates the pressure at the turbine inlet in the 2000 rpm 140 Nm with SAI engine operating point. To accurately reproduce the pressure measurement, the virtual simulation sensor is placed in the same position as the test bench. However, the absence of the turbine and wastegate in the detailed exhaust simulation compared to the test bench may cause slight differences in the pressure curves. Therefore, to minimise these differences, the boundary conditions applied upstream of the turbine are given by *QuickSim*, which utilises a 0D turbocharger model for a more precise reflection of the actual conditions in the engine.

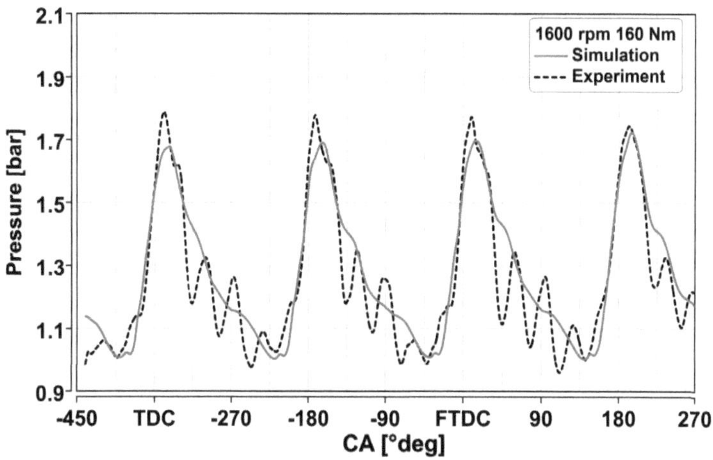

Figure 4.12: Pressure upstream of the turbine at 1600 rpm 160 Nm with SAI.

Figure 4.13 compares the temperature measurements obtained from the test bench, the 3D-CFD simulation with instant values, and the 3D-CFD simulation with the thermocouple model. The results show that incorporating the thermocouple model in the simulation significantly improves the accuracy of the temperature measurements and allows for better matching of the results to the one of test bench. This is important because post-oxidation is highly dependent on temperature, and accurate temperature measurement is crucial for precise simulation results. Understanding the limitations of measurement methods and incorporating appropriate models in simulations is essential to improve their accuracy.

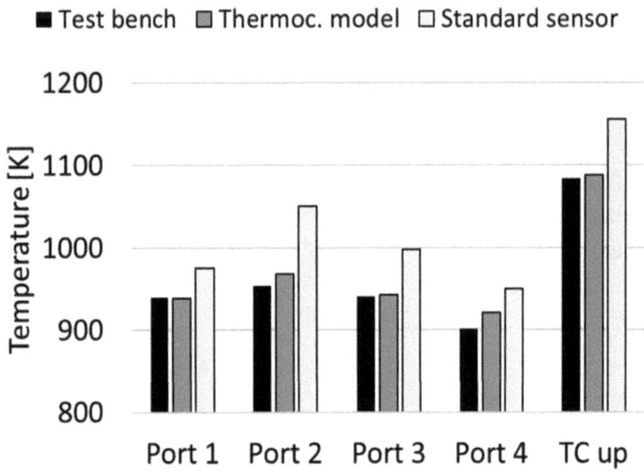

Figure 4.13: Temperature measurements at 2000 rpm 140 Nm with SAI.

5 The Full Engine Virtual Test Bench

This chapter presents the results of numerical studies at the full engine virtual test bench. After verifying the accuracy of the simulation, it becomes possible to use it to predict the most relevant phenomena that characterise the engine operating points and their post-oxidation processes. Specifically, two areas of interest have been investigated: mixture formation and inhomogeneities among the cylinders. This chapter also presents insights into different injections strategies to promote post-oxidation, such as post-injection. This secondary fuel injection can occur at different points, such as before the ignition point or before the exhaust valve opening. Understanding the parameters that trigger the post-oxidation phenomena is fundamental, as it provides the necessary foundation for investigating the post-oxidation process itself. For this reason, a dedicated study of these precursors is essential before proceeding to a detailed analysis of the post-oxidation. The results obtained from the virtual test bench enable the extraction of detailed boundary conditions at the exhaust valves, accounting for the concentration of incomplete combustion products, temperature, pressure, and mass flow in each cell during the exhaust phase. They can be applied as inputs to the detailed exhaust manifold model, where post-oxidation can subsequently be analyzed using a reaction mechanism.

5.1 Precursors of post-oxidation: mixture formation analysis

The mixture formation process within an engine is a complex phenomenon influenced by many factors, including the type of fuel utilised, the combustion chamber design and the fuel injection system. Extensive analysis has been carried out to accurately map the inhomogeneities within the cylinder, which is crucial in obtaining detailed boundary conditions for the refined exhaust model. Table 5.1 shows the lambda values for all cylinders, indicating that the inner cylinders have a richer mixture than the outer ones. This difference in lambda values can significantly impact the formation of rich combustion products. As a

M. Pipolo, *3D-CFD Virtual Development of Post-Oxidation Strategies with Secondary Air Injection for Advanced Internal Combustion Engines*, Wissenschaftliche Reihe Fahrzeugtechnik Universität Stuttgart,
https://doi.org/10.1007/978-3-658-51343-6_5

result, the oxidation of different cylinders will have distinct heat release rates, further highlighting the importance of accurately quantifying any imbalance among the cylinders. Indeed, table 5.1 clearly shows that the in-cylinder lambda values are richer during SAI operations.

Table 5.1: Lambda values for different operating conditions and cylinders.

Operating point	**Cyl. 1**	**Cyl. 2**	**Cyl. 3**	**Cyl. 4**
1600 rpm 160 Nm SAI	0.96	0.95	0.95	0.95
1600 rpm 160 Nm no-SAI	1.10	1.08	1.07	1.08
2000 rpm 140 Nm SAI	0.95	0.94	0.94	0.96
2000 rpm 140 Nm no-SAI	1.09	1.08	1.08	1.09
2000 rpm 160 Nm SAI	0.95	0.94	0.94	0.95
2000 rpm 160 Nm no-SAI	1.12	1.11	1.11	1.11

The presence of inhomogeneities results in distinct zones with significantly rich conditions, as shown in figure 5.1. The figure displays the lambda distribution for all the cylinders at the ignition point considering a side view (left side of the figure) and top one (right side of the figure) with a section through the centre of the combustion chamber for the engine operating point 1600 rpm 160 Nm operated with SAI. The same type of consideration can be applied to all other engine points. At this specific crank angle (0 °CA, FTDC), it is apparent that the mixture is incompletely homogenised, resulting in rich combustion products like CO, HC, and H_2. During combustion, the flame front may encounter volumes with a local lambda value that differs from the average value due to incomplete mixture formation.

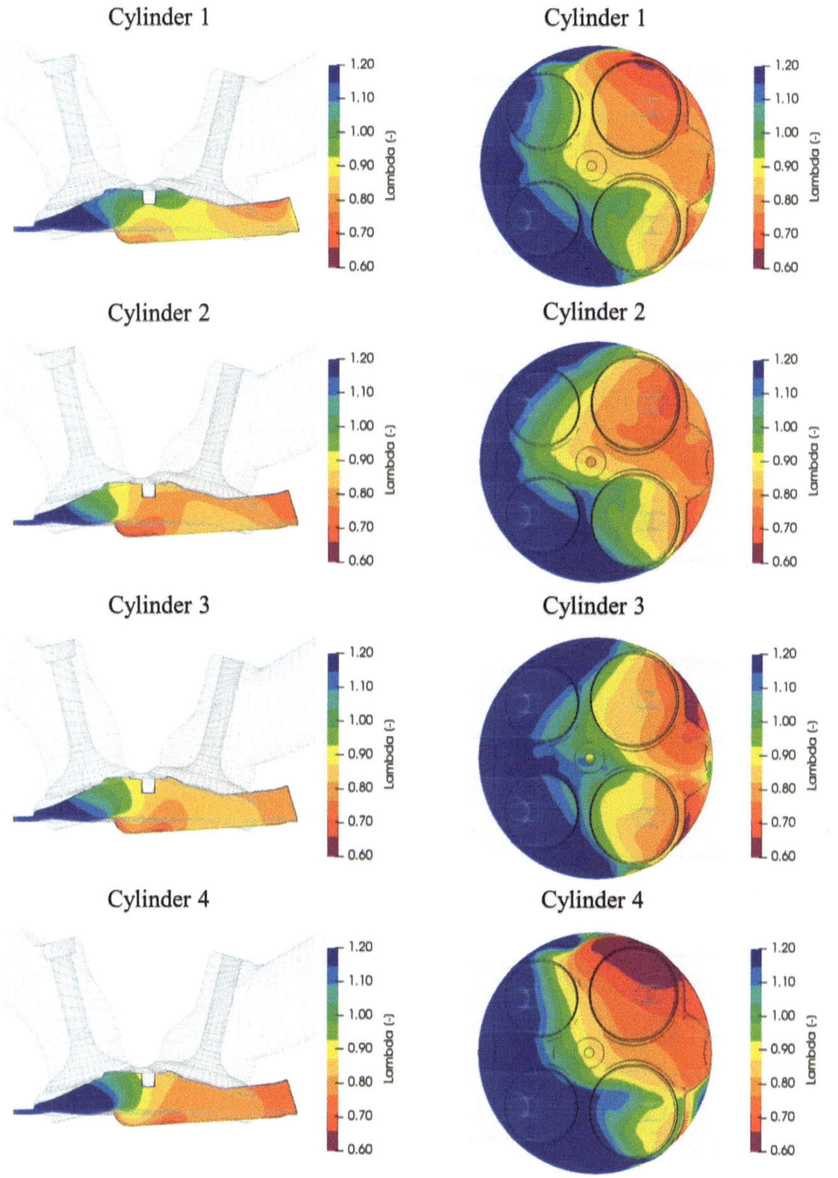

Figure 5.1: Lambda measured at IP – 1600 rpm 160 Nm with SAI.

In addition, the inhomogeneities within the cylinder contribute to different lambda distributions among the exhaust valves, shown in figure 5.2, which displays the lambda distribution for the 1600 rpm 160 Nm engine operating point with SAI during the exhaust phase (200 °CA a.FTDC) with a section (z-axis) through the middle of the cylinder. The lambda distribution shows slight variations between the two exhaust valves (the ones on the left in the figure), resulting in differences in the amounts of incomplete combustion products and reactive species discharged through each valve. Consequently, the quantity of reactive species differs across the exhaust runners, which leads to varying contributions to the post-oxidation process.

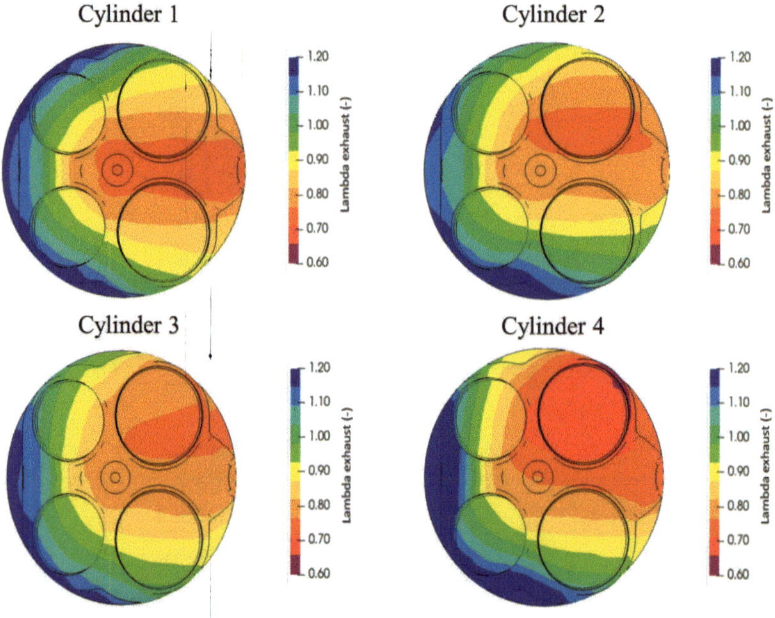

Figure 5.2: Lambda at 200 °CA a.FTDC – 1600 rpm 160 Nm with SAI.

Similar observations can also be made for the engine operating point 2000 rpm 140 Nm with SAI, shown in figure 5.3. The only notable difference between the two cases is the presence of more diffuse rich zones at 2000 rpm, which occur as the injected fuel increases.

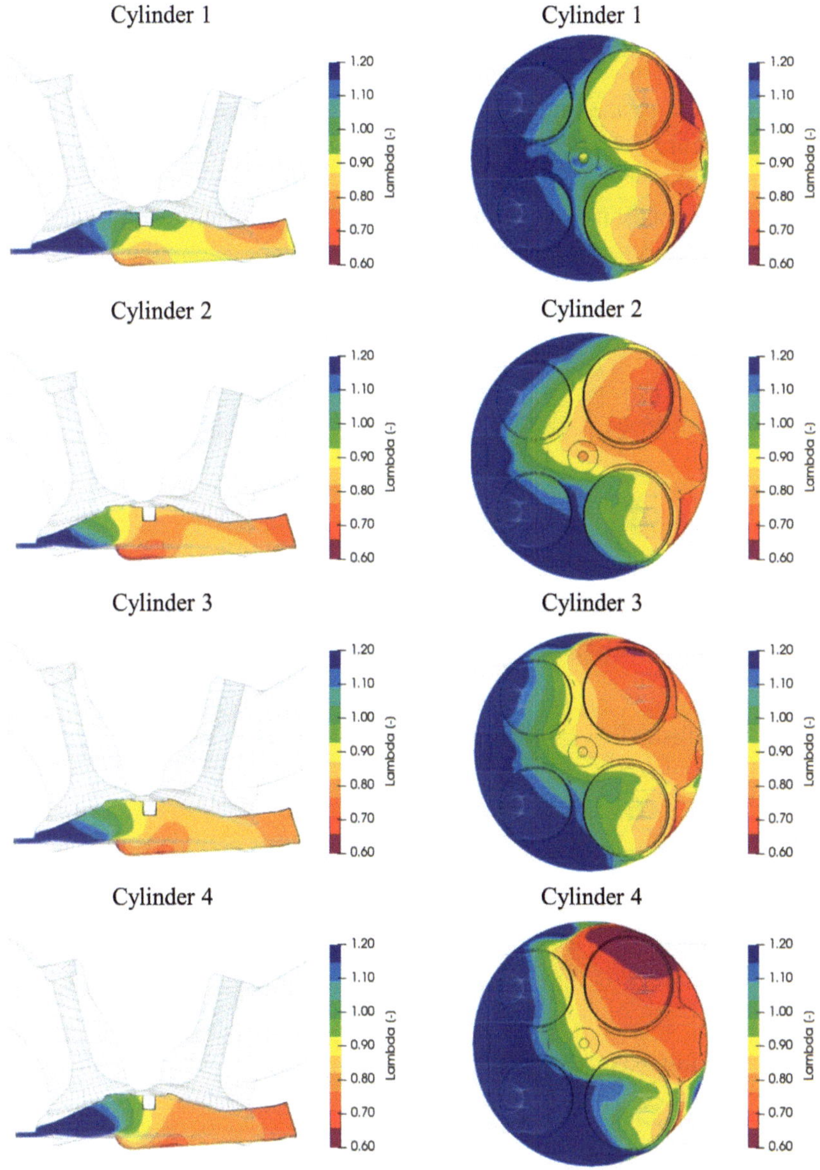

Figure 5.3: Lambda measured at IP – 2000 rpm 140 Nm with SAI.

Furthermore, when the engine is operating without SAI, i.e. at 2000 rpm 140 Nm, the quantity of injected fuel is lower, and the engine runs with a leaner lambda in the cylinder, as shown in table 4.5 (chapter 4.4). Despite the leaner lambda operation, the mixture still needs to be completely homogenised, as shown in figure 5.4. Nevertheless, rich conditions areas in the combustion chamber are still present, producing CO, H_2, and HC. These unburned products could subsequently oxidise in the exhaust system with sufficient oxygen. If secondary air injection is not operated, oxygen could be delivered to the exhaust system either by long valve overlap or by considering the oxygen present in the mixture. However, if oxygen is not provided, the oxidation process can not occur.

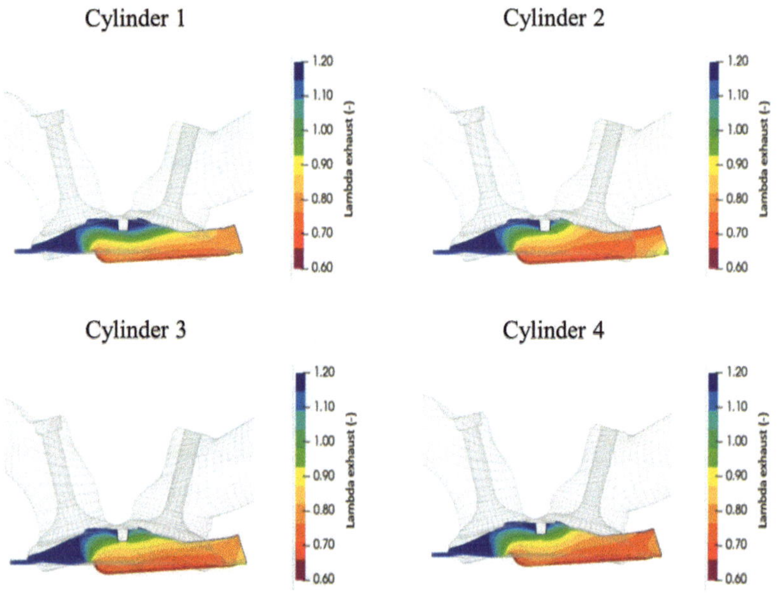

Figure 5.4: Lambda measured at IP – 2000 rpm 140 Nm without SAI.

To achieve a comprehensive insight into the in-cylinder mixture formation, the Lambda-Map tool is utilised to assess the degree of inhomogeneity present within the combustion chamber. This tool is specifically designed to monitor the progression of mixture formation throughout the compression stroke and up to the ignition point. For example, figure from 5.5 to 5.8 provide a lambda map generated for the 1600 rpm 160 Nm with SAI operating point for all the cylinders. Specifically, at each crank angle, the volume of the combustion chamber is calculated, and the air to fuel ratio is grouped into different bins within the current volume. For each crank angle then the percentage of the cylinder volume at a specific lambda can be identified. It is then useful to group the percentage of volume at each lambda into bands of multiple lambda values: the rich band (represented by a dotted line with a lambda range from 0 to 0.95), the stoichiometric band (represented by a continuous line with a lambda range from 0.95 to 1.05), and the lean band (represented by a dashed line with a lambda range from 1.05 to 3). Although the engine global lambda is 0.95, the Lambda Map analysis for cylinder 1 reveals that 50% of the combustion chamber volume exhibits a lambda richer than 0.95 at the ignition point. The percentage of this volume with a richer lambda value varies depending on the specific engine operating conditions.

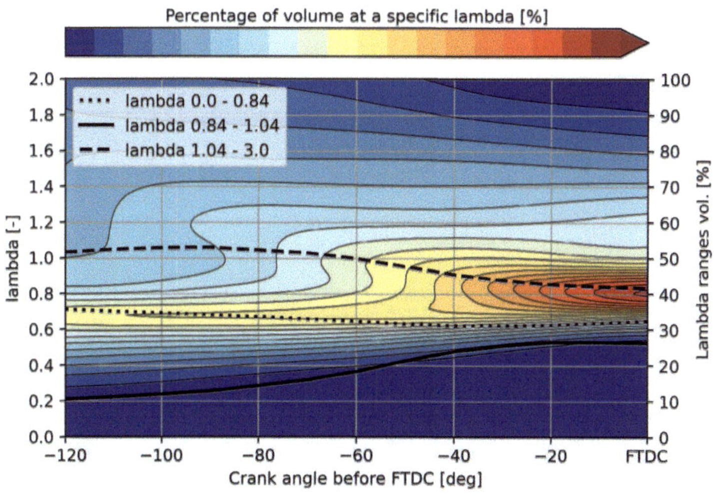

Figure 5.5: Lambda map – 1600 rpm 160 Nm with SAI, cylinder 1.

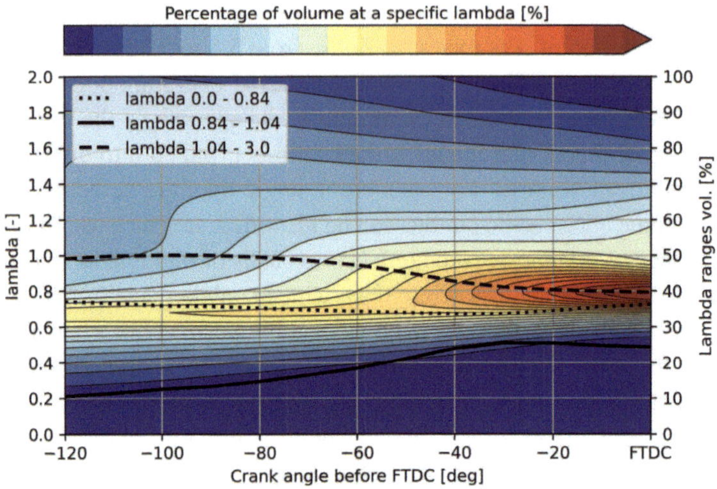

Figure 5.6: Lambda map – 1600 rpm 160 Nm with SAI, cylinder 2.

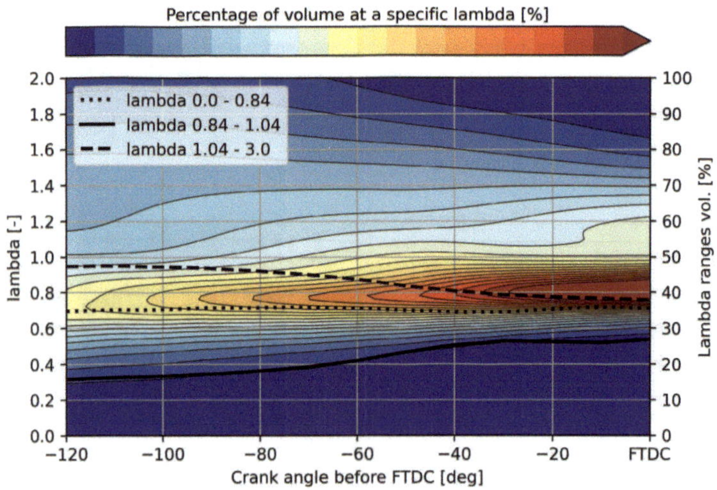

Figure 5.7: Lambda map – 1600 rpm 160 Nm with SAI, cylinder 3.

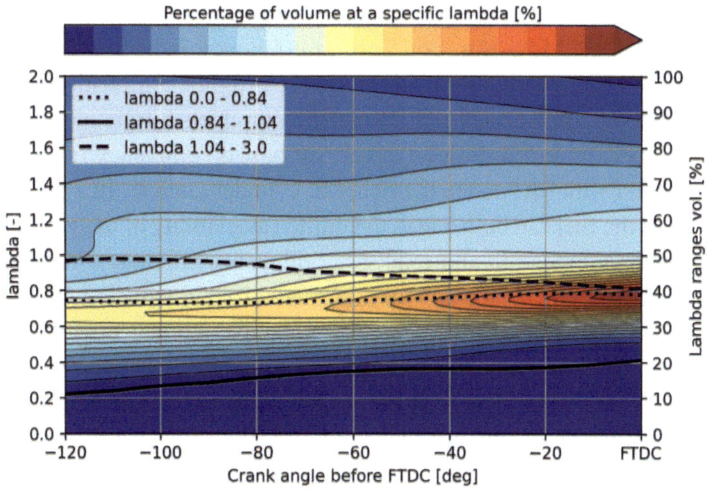

Figure 5.8: Lambda map – 1600 rpm 160 Nm with SAI, cylinder 4.

5.2 Overall engine behaviour: imbalance among the cylinders

Assessing the imbalance among the cylinders is also important, possibly due to various factors, such as engine design, location and geometry of the intake and exhaust ports, and fuel delivery systems. 3D-CFD simulations are crucial in quantifying these differences by simulating the flow and combustion in each cylinder. The values presented in Table 5.2 (no-SAI) and Table 5.3 (SAI 16%) for each cylinder at 2000 rpm 140 Nm provide crucial insights into the formation of rich combustion products. The combustion process in an engine is a complex phenomenon that involves various parameters and factors, including the timing and duration of combustion, combustion efficiency, EGR and lambda. These parameters can significantly impact the formation of pollutants and affect engine performance. In particular, the mass fraction burned 50 (MFB50) and mass fraction burned 10-90 (MFB10-90) are crucial indicators that provide insights into the timing and duration of combustion. There are slight differences in MFB50 and MFB10-90 values among the cylinders for both engine operating points. The engine operating point with SAI exhibits a faster combustion rate than without SAI. Moreover, combustion efficiency indicates the degree to which the fuel is converted into useful work. A higher combustion efficiency would reduce the formation of rich combustion products by completely burning the fuel. Hence, table 5.3 shows a lower combustion efficiency for the engine operating point with SAI due to the richer in-cylinder lambda. In a rich mixture, only some of the fuel can burn due to the limited availability of oxygen, resulting in incomplete combustion and reduced combustion efficiency. Additionally, the recirculation of exhaust gases plays a crucial role in reducing emissions by lowering the combustion temperature and reducing the production of pollutants. The table shows that the EGR percentages for the internal cylinder are higher than those for the outer one, which could be attributed to the engine design, particularly the location and geometry of the intake and exhaust ports.

Table 5.2: Combustion parameters, no-SAI application.

Parameter	**Cyl. 1**	**Cyl. 2**	**Cyl. 3**	**Cyl. 4**
MFB50, °CA a.FTDC	33.1	33.5	33.3	32.8
MFB10-90, °CA	27.5	28.5	28.3	27.9
Comb. Eff., %	92.4	90.1	90.6	90.8
EGR, %	5.25	5.8	5.7	5.2

Table 5.3: Combustion parameters, application with SAI 16%.

Parameter	**Cyl. 1**	**Cyl. 2**	**Cyl. 3**	**Cyl. 4**
MFB50, °CA a.FTDC	30.1	30.7	30.7	29.5
MFB10-90, °CA	26.8	27.8	27.3	26.5
Comb. Eff., %	88.4	88.4	88.2	86.7
EGR, %	5.5	6.2	5.9	5.4

Moreover, by analysing the in-cylinder mass and in-cylinder lambda at different speeds, it is possible to identify which cylinders take in more air and how this affects the resulting combustion process. Figure 5.9 depicts the differences among the cylinders based on in-cylinder mass and in-cylinder lambda for 1600 rpm (upper part of the figure) and 2000 rpm (lower part of the figure) with SAI.

At 1600 rpm, cylinder 1 has more air and, therefore, a leaner lambda, while at 2000 rpm, both cylinder 1 and 4 (outer cylinders) have more air and a leaner lambda compared to the inner cylinders.

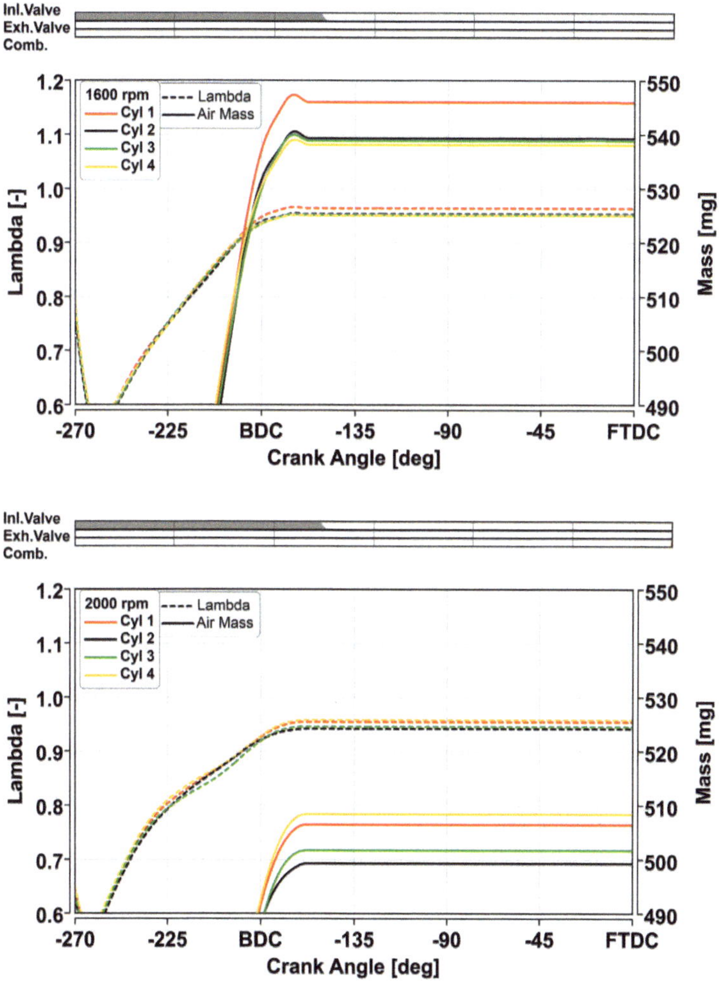

Figure 5.9: Lambda and air mass at 1600 rpm and 2000 rpm.

5.3 Injection strategies for post-oxidation

Another post-oxidation strategy analysed during the project involves secondary fuel injection within the cylinder and SAI in the exhaust manifold. The primary direct injection occurs during the intake stroke, while the secondary injection is set to take place during the compression, power, or exhaust stroke. It is assumed that this post-injection and SAI can assist with post-oxidation phenomena in the exhaust manifold, as fuel is injected during the late compression, power, or exhaust stroke, thereby increasing the number of rich combustion products. These gases can then react with oxygen from the secondary air injectors. The timing of the post-injection is a critical factor in post-oxidation, directly affecting emissions. Achieving efficient post-oxidation and minimising emissions requires optimising the timing, the ratio between primary and post-injection, and the appropriate amount of SAI. Additionally, the oxidation reactions require sufficient time and energy to occur. If fuel is injected too late in the exhaust stroke, the post-oxidation process may not initiate effectively due to the lack of energy or time for chain-branching reactions. Within the 3D virtual environment, two cases of secondary fuel injection were considered at 1600 rpm 150 Nm: secondary injection at 710 °CA (10 °CA b.FTDC), also known as late-injection, and secondary injection during the exhaust phase at crank angle 880 °CA (160 °CA a.FTDC), also known as post-injection. In figure 5.10 and figure 5.11, a comparison of in-cylinder turbulence and the lambda is presented. The graphs show that in the case of late-injection (710 °CA), the lambda in the cylinder is richer before combustion. This indicates a higher fuel concentration relative to air, which can enhance the combustion process but may also increase the risk of PM emissions. The secondary fuel injection occurs just 10 crank angles before the ignition point, leading to a peak in turbulent kinetic energy. This peak in TKE is crucial as it enhances the mixing of fuel and air, promoting a more efficient combustion process.

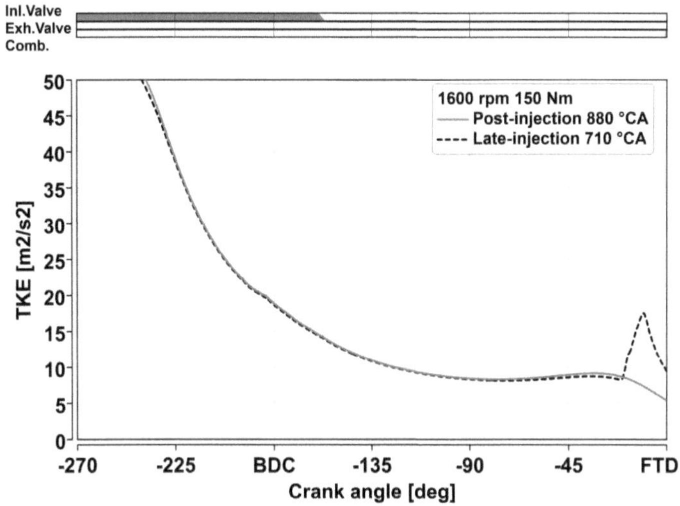

Figure 5.10: Turbulence for late and post-injection.

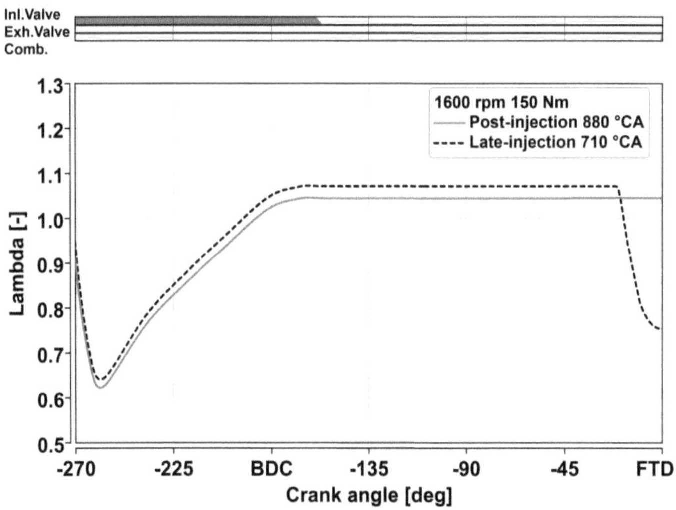

Figure 5.11: Lambda for late and post-injection.

Additionally, figure 5.12 and figure 5.13 compare the temperature at the spark and the heat release rate between the injection strategies. For the late-injection scenario, a lower temperature at the spark is observed. This temperature reduction is due to fuel evaporation, which cools the in-cylinder temperatures. Hence, as the exhaust valve opens, the trends for the two cases invert. In the post-injection case, a temperature drop is noted due to the fuel evaporation occurring shortly after the exhaust valve opens. The heat release rate also differs between the two scenarios. Initially, the combustion process is slower in the late-injection case and only becomes faster after reaching MFB50. It is important to note that both cases show the same amount of heat released by the end of the combustion process despite the late-injection case using a more significant amount of fuel. This indicates that during late-injection, part of the fuel is crossed by the flame front but does not entirely burn due to locally rich fuel zones. This results in the formation of highly reactive incomplete combustion products, which still have significant energy potential. These products react later in the exhaust system through interaction with oxygen.

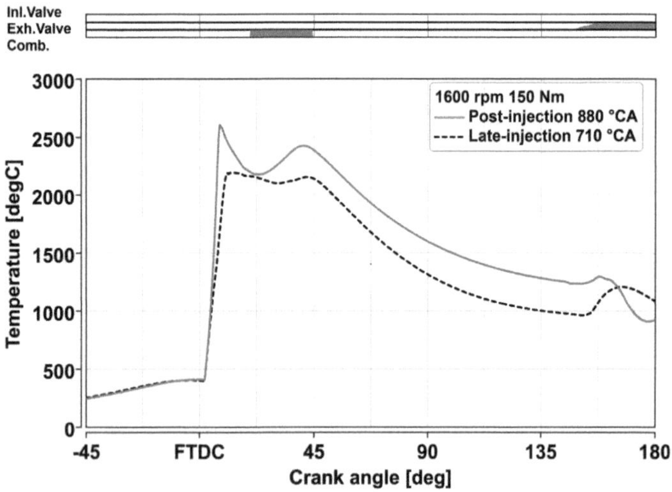

Figure 5.12: Temperature at the spark plug for late and post-injection.

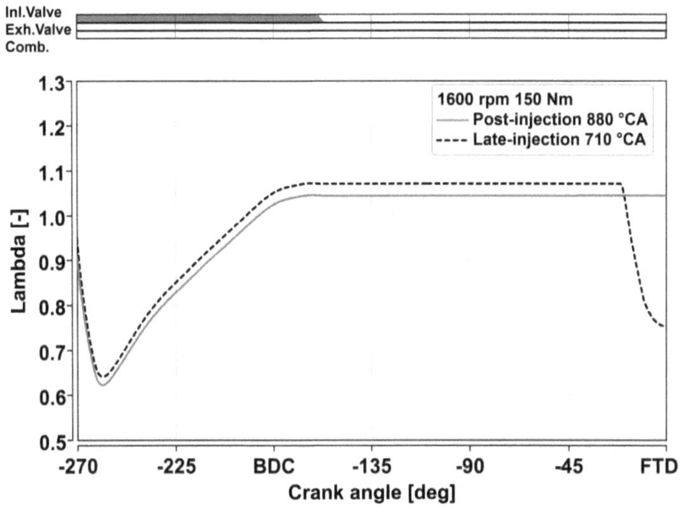

Figure 5.13: HRR for late and post-injection.

The complete engine analysis can be further refined by examining the following image, which illustrates the lambda distribution inside the cylinder between the cases late-injection and post-injection. This is depicted using a clip volume with a cross-section passing through the centre of the cylinder. Figure 5.14 (upper part of the figure) shows the lambda at 710 °CA (10 °CA b.FTDC). Both late and post-injection cases have a non-homogeneous mixture, with lean zones near the exhaust valve and rich zones near the intake valve (injector side). This leads to the formation of locally rich zones and, consequently, a non-optimal combustion process with many incomplete combustion products. This is particularly evident in the late-injection case on the left side of the figure due to the secondary fuel injection close to the IP. Furthermore, the exact figure (lower part of the figure) shows the lambda at 880 °CA (160 °CA b.FTDC), just before the exhaust valves open. During the post-injection (right side), the fuel evaporates in the cylinder and is discharged upon the exhaust valve opening. Hence, in this case, there will be a high concentration of THC in the exhaust, which requires energy and time to burn.

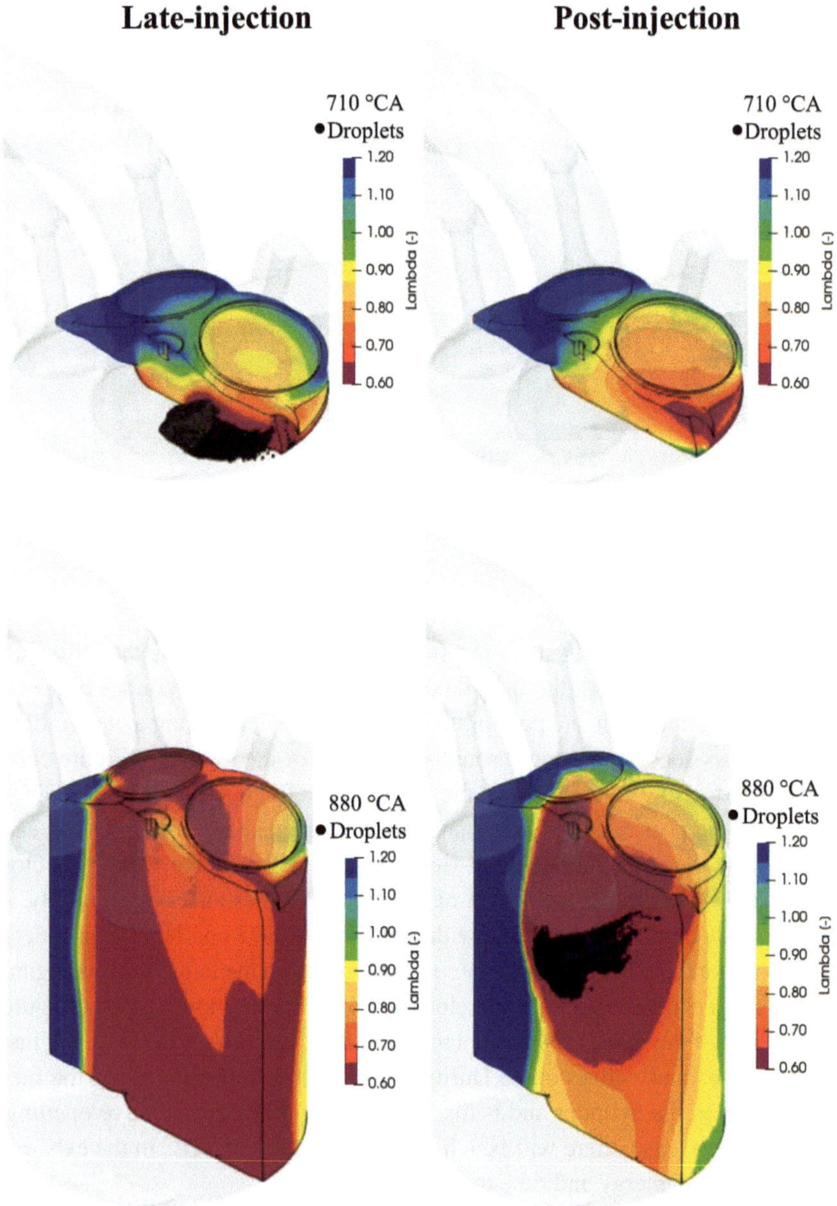

Figure 5.14: Lambda at different time step with late and post injection.

6 Detailed Exhaust Model and Post-Oxidation Strategies

After validating the 3D-CFD full engine simulations, the information needed to set the input boundary conditions of the numerical model was collected. This included data on temperature, pressure, mass flow, and mass fraction of chemical species. The fine-mesh model describes the physical and chemical phenomena occurring in the exhaust manifold during the engine cycle and identifies the variables that most significantly influence post-oxidation. The simulated volume is confined to the exhaust manifold, requiring a significantly refined mesh built using the CFD software *StarCCM+*. This software can implement and easily modify a chemical reaction mechanism. The simulation reproduces the experimental methodology, including the Mexa 1600 gas analyser, which measures various emissions such as CO, CO_2, O_2, and HC. The computational mesh consists of the exhaust manifold from the exhaust valves to the turbine inlet, including the 10 cm adaptor between the cylinder head and the four-in-one exhaust manifold, as shown in chapter 4, figure 4.6. The boundary conditions given to *StarCCM+* are evaluated in the 3D-CFD tool *QuickSim* at the exhaust valves, taking into account the mass of rich combustion products, temperature, pressure, and mass flow. Consequently, the concentrations are extrapolated, and a complete set of detailed boundary conditions is established. This chapter offers a detailed analysis of post-oxidation in the original exhaust system utilizing secondary air injection. The findings provide a reference point for evaluating the effectiveness of post-oxidation and identifying the key factors that influence it. Once this reference was established, additional studies were undertaken to enhance the process. These studies included exploring various exhaust geometries to improve the mixing of combusted gases with oxygen and analysing injection strategies for post-injection to increase the number of reactive species.

M. Pipolo, *3D-CFD Virtual Development of Post-Oxidation Strategies with Secondary Air Injection for Advanced Internal Combustion Engines*, Wissenschaftliche Reihe Fahrzeugtechnik Universität Stuttgart, https://doi.org/10.1007/978-3-658-51343-6_6

6.1 Application of secondary air injection

The outcomes of the secondary air injection applied with the original geometry can be summarised in the following case studies:

1. 2000 rpm 140 Nm with and without SAI;
2. 1600 rpm 160 Nm and 2000 rpm 140 Nm with SAI;
3. 1600 rpm and 2000 rpm with increased SAI temperature.

Case 1 examines the effects of operating the same engine point with and without secondary air injection. Case 2 looks at two different engine points, both equipped with SAI. In case 3, the same engine points from Case 2 are tested with a higher secondary air temperature to determine whether there are significant increases in post-oxidation efficiency.

6.1.1 Case 1: SAI vs. no-SAI

The primary indication for evaluating the effectiveness of post-oxidation is the overall heat release rate of the chemical reactions, as illustrated in Figure 6.1 for both engine operating conditions. In the case of the SAI application (represented by the black line), post-oxidation takes place in the exhaust manifold. This process is more significant for the inner cylinders, specifically cylinders 2 and 3, compared to the outer cylinders. This difference can be attributed to the varying lambda values observed between the cylinders, as reported in Table 5.1 in chapter 5.1. These variations are the result of cylinder-to-cylinder differences and inhomogeneities, leading to a greater presence of incomplete combustion products, which can be utilized as fuel for post-oxidation. The heat released during the chemical reactions increases the temperature of the exhaust gas, facilitating the heating of the TWC. Without SAI, the heat released from these chemical reactions is significantly low, leading to ineffective post-oxidation. This ineffectiveness occurs mainly because the engine runs on a lean mixture, which limits the availability of rich combustion products that could be used as fuels for post-oxidation. The low heat release observed in this engine condition is primarily due to inhomogeneities within the cylinder, despite the lean lambda operation, leading to a low concentration of rich combustion products.

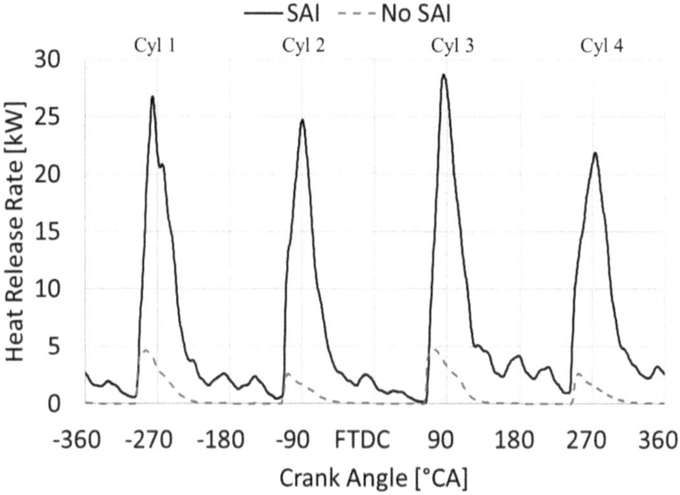

Figure 6.1: HRR for SAI and without SAI (x-axis refers to cyl. 1).

Figure 6.2 illustrates the chemical heat release rate and temperature at the exhaust of cylinder 3 (270 °CA a.FTDC) for both engine operating points. Initially, the hot exhaust stream encounters a cloud of fresh air injected by the SAI system next to the exhaust valve. This leads to challenges in mixing due to the differences in their densities. The first indications of post-oxidation can be observed in the adaptor near the air injector, where air is continuously injected during the exhaust phase. Most of the heat release occurs before reaching the turbine, where improved mixing conditions and higher temperatures are present.

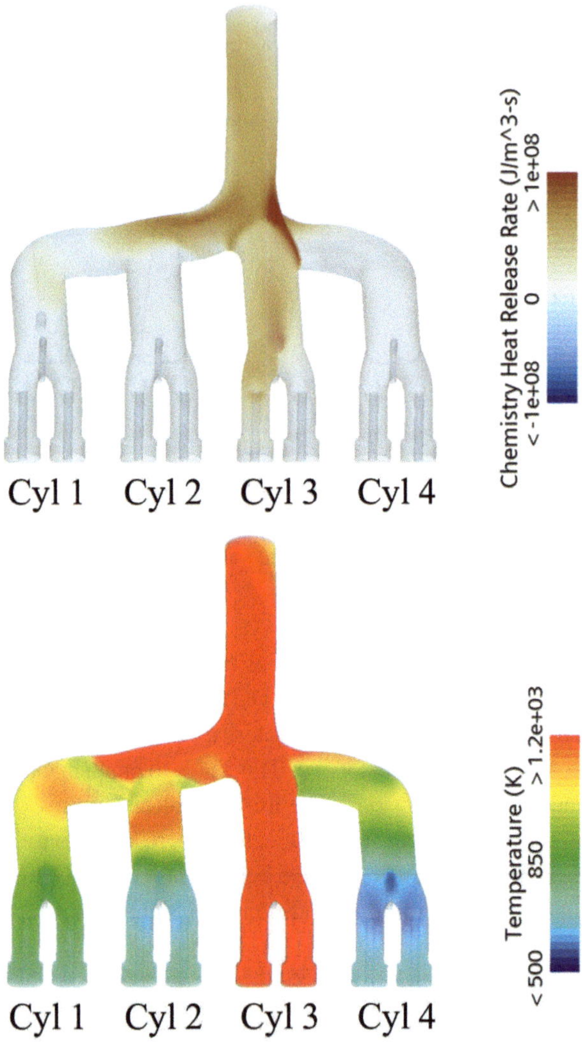

Figure 6.2: HRR and temperature during the exhaust of cyl. 3 (x-axis refers to cyl. 1).

The validation process focuses on confirming the molar concentrations of the most relevant species. At the test bench, a Mexa 1600 gas analyser operates with a constant mass flow rate to measure pollutant concentrations in exhaust gases. In contrast, the 3D-CFD simulation calculates species concentrations based on chemical reaction mechanisms, which may not fully capture the complexities present in the actual system. To accurately replicate the test bench methodology in the simulation, it was crucial to reproduce the sampling of species at a constant mass flow rate, similar to that of the Mexa 1600 system. Table 6.1 compares the molar concentrations of CO, CO_2, and O_2 measured by the Mexa, both in the simulation and at the test bench for the SAI application. The results show a good agreement between the experiment and the numerical environment for CO_2 concentrations at both measured positions. The observed decrease in oxygen concentration indicates that oxygen is being consumed, while the reduction in CO concentration suggests that CO is being further oxidized.

Table 6.1: Composition at adaptor 1 and turbine inlet for SAI application.

Mol fraction (mol/mol)%	Adaptor 1		Turbine inlet	
	Exp.	Sim.	Exp.	Sim.
O_2	4.83	5.35	2.18	3.19
CO_2	10.34	10.41	12.6	12.27
CO	2.96	1.49	2.39	0.66

Additionally, figures 6.3 and 6.4 illustrate the mole fractions of CO and O_2 during the exhaust phase of cylinder 3 (270 °CA a.FTDC), both with and without SAI. These figures show multiple cross-sections through the middle of the exhaust. In the scenario with SAI, the left exhaust valve discharges a higher concentration of CO compared to the right valve, due to inhomogeneities within the cylinder. It is also noticeable that some CO remains near the ports of the cylinders that have already discharged. In contrast, when SAI is not utilised, the CO mole fraction in the exhaust is significantly lower. Moreover, when secondary air injection is implemented, air clouds (visible in the bottom part of the figure 6.3) can be observed near the exhaust ports. Without SAI, the oxygen content in the exhaust gases primarily comes from the limited valve overlap and the lean fuel mixture. As a result, the O_2 mole fraction in the exhaust manifold is lower compared to the scenario where SAI is used.

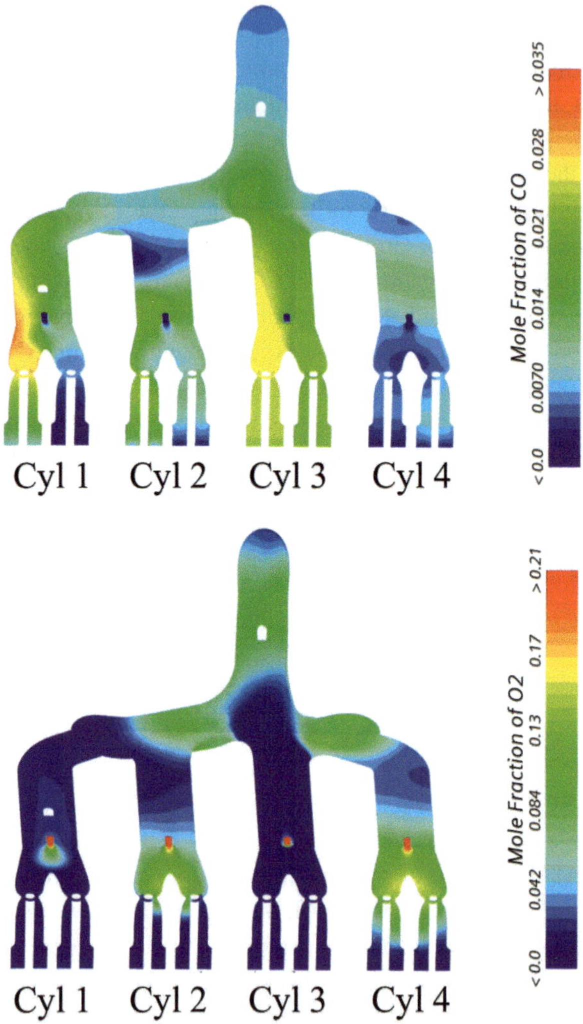

Figure 6.3: Mole fraction of CO and O_2 for SAI application.

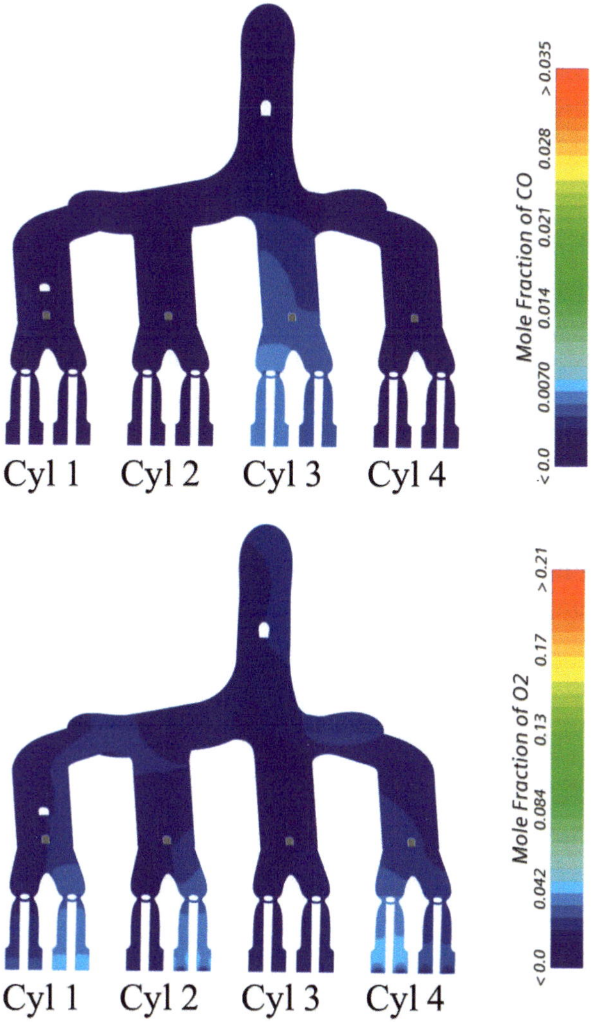

Figure 6.4: Mole fraction of CO and O_2 without SAI.

The SAI operation shows significant post-oxidation. Introducing fresh air in the exhaust manifold enhances mixing and oxidation, which increases the heat release rate and the exhaust gas temperature. When SAI is not used, the lean fuel mixture reduces the availability of oxidizable products, resulting in a less effective post-oxidation. Once the benefits of SAI in promoting post-oxidation have been established, the next step is to investigate its behaviour under different operating conditions.

6.1.2 Case 2: SAI at different operating conditions

This case illustrates the use of SAI at two specific engine operating points: 1600 rpm 160 Nm and 2000 rpm 140 Nm. Figure 6.5 displays the overall HRR of oxidation reactions for both engine operating points. The inner cylinders exhibit a greater post-oxidation rate than the outer cylinders. This can be attributed to the observed differences in lambda values between the cylinders caused by cylinder-to-cylinder variations, shown in chapter 5.1. As a result, there are more incomplete combustion products available to participate in post-oxidation. Additionally, the heat release rate for the 2000 rpm case is faster and more significant than that of the 1600 rpm case, resulting in more energy being released during one engine cycle.

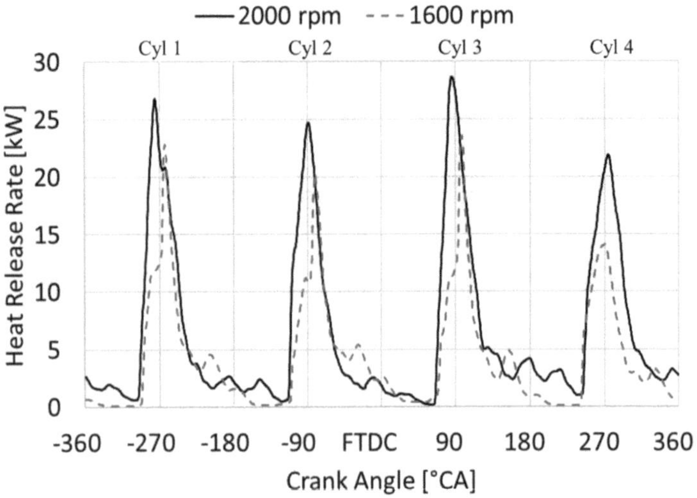

Figure 6.5: HRR for 2000 and 1600 rpm with SAI (x-axis refers to cyl. 1).

Measuring the energy released per cycle is an effective way to assess the efficiency of oxidation reactions. To evaluate the contribution of different areas to post-oxidation, the detailed mesh of the exhaust manifold can be divided into distinct regions. By examining the adaptors and the exhaust manifold (up to the turbine inlet) separately, it is possible to identify where the most post-oxidation occurs. Table 6.2 presents the energy released per cycle and the percentage contribution of each region to the total post-oxidation energy for both engine operating points. Most of the energy is released upstream of the turbine, confirming the better mixing conditions compared to the adaptors. At 2000 rpm, the turbine releases 57.1% of the total post-oxidation energy, approximately 10 percentage points more than at 1600 rpm (47.2%), which can be attributed to the higher amount of fuel injected into the cylinder. This leads to more incomplete combustion products, enhancing post-oxidation. In contrast, the energy released in the adaptors remains constant between 1600 rpm and 2000 rpm, with differences of less than 4 J/cyc in most cases. However, their percentage contribution to total post-oxidation energy decreases at higher rpm. For example, Adaptor 1 contribution drops from 13.5% at 1600 rpm to 9.8% at 2000 rpm, reflecting the greater contribution to post-oxidation of the turbine region at higher speeds. The observed trends can be attributed to the higher temperatures of the turbine, which is continuously exposed to exhaust gases, thus promoting oxidation. On the other hand, the adaptors are continuously cooled by the fresh secondary air, which is injected throughout the engine cycle, thereby reducing the effectiveness of post-oxidation in these regions.

Table 6.2: Energy released in different region for SAI applications.

Region	**1600 rpm 160 Nm**		**2000 rpm 140 Nm**	
	J/cyc	**Energy released, %**	**J/cyc**	**Energy released, %**
Adaptor 1	44	13.5	40	9.8
Adaptor 2	40	12.4	52	12.8
Adaptor 3	45	13.8	40	9.8
Adaptor 4	43	13.1	42	10.4
Turbine	154	47.2	232	57.1
Total	326	100	406	100

A critical parameter for defining optimal conditions for post-oxidation is the λ_{CO} value. This value provides insight into the necessary conditions for the oxidation of carbon monoxide to occur at the equilibrium, indicating whether there is an adequate concentration of both CO and oxygen O_2 to sustain post-oxidation. A λ_{CO} value less than 1 signifies a rich condition, meaning there is more CO present than required. On the other hand, a value greater than 1 indicates a lean condition, where there is less CO available than necessary to react with O_2. It is essential to understand that λ_{CO} does not represent the actual lambda value of the engine. Equation eq. 6.1 defines how λ_{CO} is calculated:

$$\lambda_{CO} = \frac{\alpha_{\mathrm{CO}}}{\alpha_{\mathrm{st}}} = \frac{\left(\frac{w_{O_2}}{w_{CO}}\right)}{\alpha_{\mathrm{st}}} \qquad \text{eq. 6.1}$$

where w_{CO} and w_{O2} represent the concentrations of carbon monoxide and oxygen, α_{st} is the stoichiometric ratio required for the complete oxidation of CO to CO_2. Figure 6.6 shows the chemical heat release rate and λ_{CO} during the exhaust phase of cylinder 3, while figure 6.7 refers to cylinder 2, both at 1600 rpm. Due to inhomogeneities within the cylinder already observed, the mole fraction of CO exhausted by the exhaust valves differs, resulting in the right exhaust valve being more reactive than the left one. Furthermore, λ_{CO} values closer to 1 lead to more effective post-oxidation.

Operating with SAI at 1600 rpm and 2000 rpm leads to higher post-oxidation rates in the inner cylinders due to the richer lambda conditions, resulting in more incomplete combustion products for oxidation. At 2000 rpm, increased in-cylinder fuel injection improves oxidation, as seen by a higher heat release rate in the turbine. The area upstream of the turbine contributes more to post-oxidation, while the adaptors contribution decreases at higher rpm due to cooling from the secondary air. The λ_{CO} value confirms that post-oxidation is more effective when CO and oxygen are better balanced. Following the benefits of SAI in promoting post-oxidation, the next chapter explores the impact of increasing the injected air temperature on heat release. This approach aims to improve the energy released per cycle by oxidation reactions without major changes to the system.

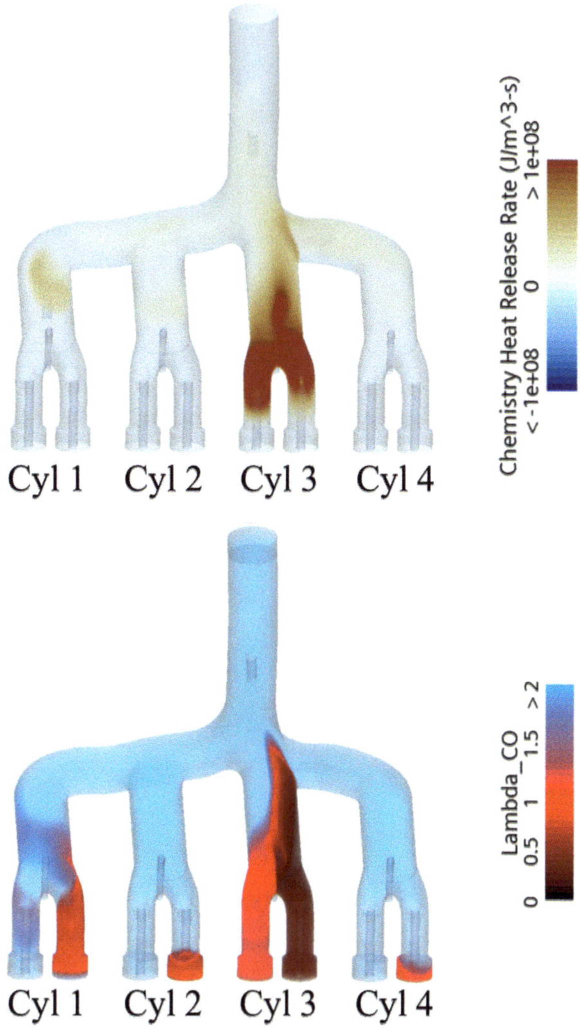

Figure 6.6: HRR and λ_{CO} during the exhaust of cyl. 3 at 1600 rpm.

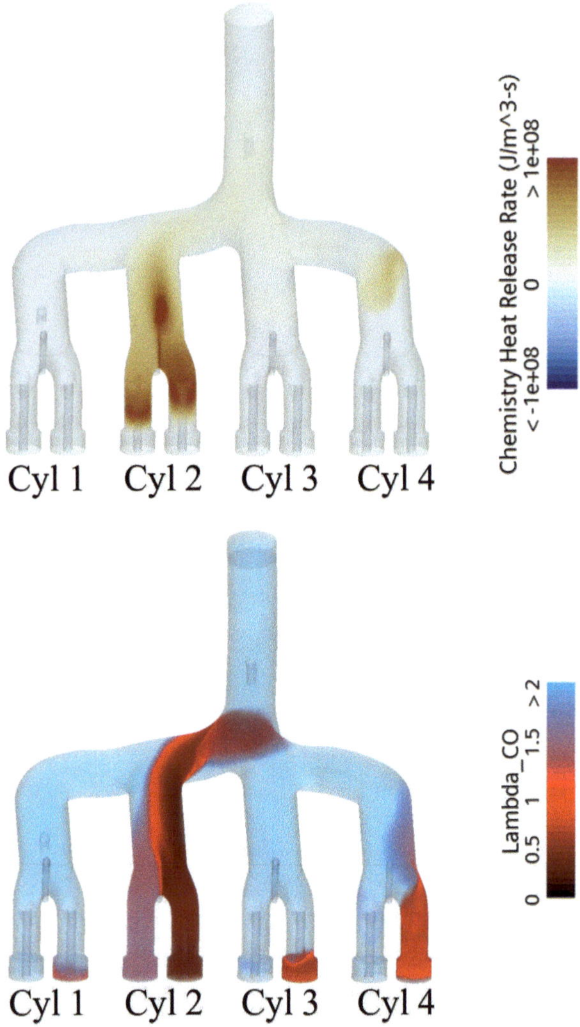

Figure 6.7: HRR and λ_{CO} during the exhaust of cyl. 2 at 1600 rpm.

6.1.3 Case 3: increased SAI temperature

Investigating the effect of SAI temperature on post-oxidation is a critical aspect of exploring potential improvements. To study this effect, both engine operating points were run with increased SAI temperature, from 300K (case 2) to 323K, corresponding to a 7.5% increase. This value was chosen because it represents an actual temperature that can be found in practical applications. It is worth mentioning that the volumetric flow rate of secondary air was kept constant during the test. Table 6.3 assesses the impact of the SAI temperature on post oxidation, showing the energy released per cycle for each engine region and considering the ΔEnergy with case 2 (reference case, table 6.2). The enhancement achieved through increasing the SAI temperature was particularly significant at 1600 rpm, where an improvement in the energy released was observed in each region. Overall, the total energy released increased by 5% compared to reference case. On the other hand, at 2000 rpm, the improvement was only 1%, and only adaptor 4 had an higher percentage than the standard case. At 1600 rpm, due to the longer engine cycle, the exhaust gases spend more time in the exhaust system, giving more time at the post-oxidation reactions to occur. As the temperature of SAI increases, the thermal energy of oxygen molecules also increases, resulting in more oxygen molecules having enough energy to overcome the activation energy barrier required for the reaction, enhancing the post-oxidation.

Table 6.3: Energy variation with increased SAI temperature.

Region	**1600 rpm**		**2000 rpm**	
	J/cyc	**ΔE with case 2, %**	**J/cyc**	**ΔE with case 2, %**
Adaptor 1	46	+4.5%	40	0%
Adaptor 2	41	+2.5%	52	0%
Adaptor 3	46	+2.2%	40	0%
Adaptor 4	45	+4.6%	43	+2.3%
Turbine	163	+5.8%	234	+0.9%
Total	342	+5%	410	+1%

Figure 6.8 illustrates the temperature measurement at the 1600 rpm 160 Nm engine point for cases 2 and 3 using the thermocouple model. It is noted that as the temperature of the secondary air increases, on the one hand, post-oxidation

is slightly increased, and on the other hand, the cooling effect of the injected air is reduced since the latter is injected continuously throughout the operating cycle.

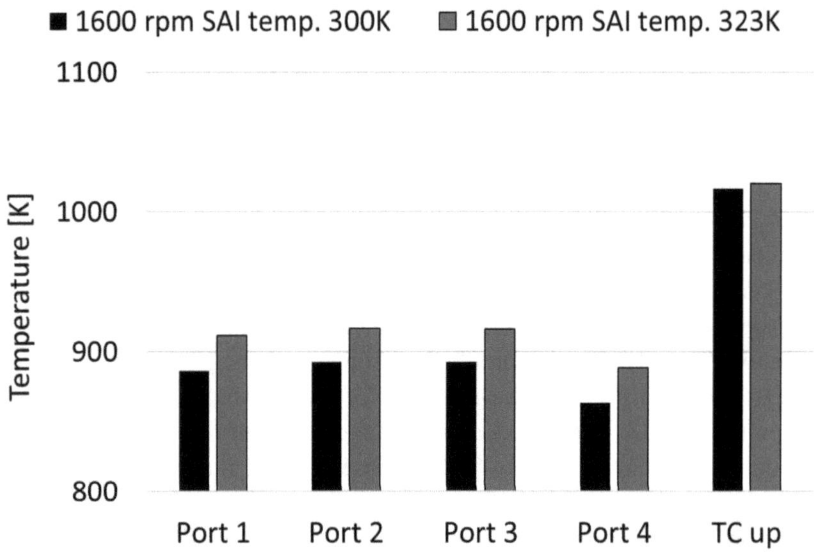

Figure 6.8: Temperature comparison between case 2 and case 3 at 1600 rpm.

The increase in post-oxidation with higher SAI temperature is driven by thermodynamic and kinetic effects. Warmer secondary air raises the thermal energy of oxygen molecules, increasing the fraction that can overcome the activation energy barrier, thus accelerating oxidation. At 1600 rpm, the longer residence time of exhaust gases allows oxidation to proceed further, explaining the greater energy release compared to 2000 rpm, where shorter residence time limits the effect. Additionally, while higher SAI temperature enhances oxidation, it also reduces the cooling effect of injected air, maintaining a more favourable thermal environment for post-oxidation reactions. Having established a benchmark for heat release across various case studies, the next step is to investigate whether the process can be further enhanced by modifying the exhaust manifold design.

6.2 Innovative exhaust manifold designs

Previous studies have shown that mixing unburned gas species and oxygen in the exhaust pipe significantly affects post-oxidation [37, 38, 39, 40]. Hence, the exhaust manifold was modified with two configurations of internal bypass adaptor, as illustrated in 6.9. This design features a 20 mm diameter internal hole on the side of each exhaust line. This hole allows exhaust gases from the exhaust port to flow into other exhaust lines. Internal bypass adaptor #1 bypasses all exhaust lines, while internal bypass adaptor #2 separately bypasses exhaust ports 1 and 2, as well as exhaust ports 3 and 4. Internal #2 is expected to prevent heat loss due to bypassing, i.e., lower exhaust gas temperatures, while maintaining the mixing effect.

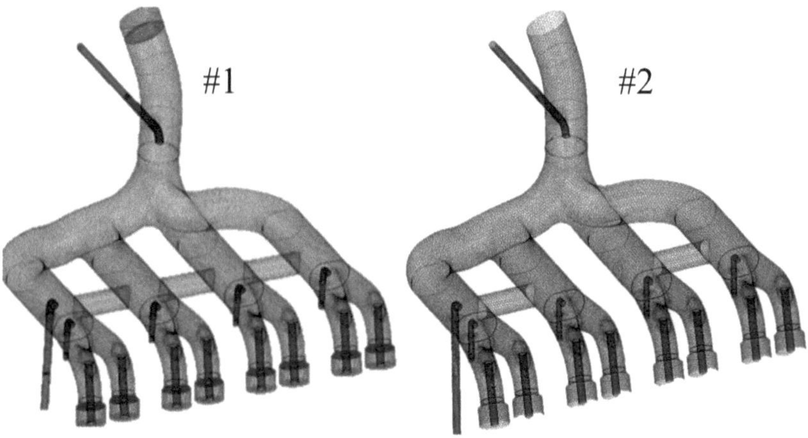

Figure 6.9: New exhaust manifold designs.

6.2.1 Internal bypass adaptor #1

To evaluate the potential of the new exhaust design, the same engine operating points used for the conventional geometry were considered, 1600 rpm and 2000 rpm, as previously discussed for cases 1 and 2. These operating points show that the majority of heat release from oxidation reactions occurs in the manifold, where optimal mixing conditions are present. The internal bypass adaptors aim to enhance mixing in the exhaust runners by allowing oxygen to flow freely through the adaptors. In the current study, the SAI strategy has been maintained as in the conventional geometry. A preliminary indication of the higher oxygen concentration within the internal bypass adaptor #1 can be observed in Table 6.4, which compares the molar concentration of O_2 measured upstream of the turbine at 1600 rpm and 2000 rpm. At both engine operating points, a significant increase in the molar concentration of O_2 is observed when using the bypass adaptor #1. Specifically, at 1600 rpm, the oxygen concentration increases from 2.8% (conventional) to 3.3% (bypass adaptor #1), showing a 17% increase. At 2000 rpm, the concentration rises from 2.1% (conventional) to 2.5% (bypass adaptor #1), a 19% increase. The higher oxygen concentration suggests a decrease in post-oxidation efficiency, which may indicate the need for further optimization in the exhaust system design.

Table 6.4: Mol fraction of O_2 for different geometries at TC up.

Mole fr. of O_2 (mol/mol)%	**1600 rpm**	**2000 rpm**
Conventional	2.8	2.1
Bypass adaptor #1	3.3	2.5

Multiple cross-sections through the middle of the exhaust have been taken to investigate the mole fraction of O_2, providing a clearer view of the oxygen concentration throughout the system. The results have been presented in figure 6.10, which illustrates the mole fraction of oxygen with the conventional geometry and the internal bypass adaptor #1 during the exhaust of cylinder 3 (270 °CA a.FTDC). In the initial phase of exhaust, oxygen is first directed to the outer adapters before returning to the inner ones. Afterwards, the oxygen mixes with the exhaust gas stream that is discharged from cylinder 3. However, if there is too much oxygen, it can cause an unexpectedly strong cooling effect, which may reduce the post-oxidation efficiency. This observation is discussed in chapter 6.1.3 and is supported by the findings of Serrano et al. [57].

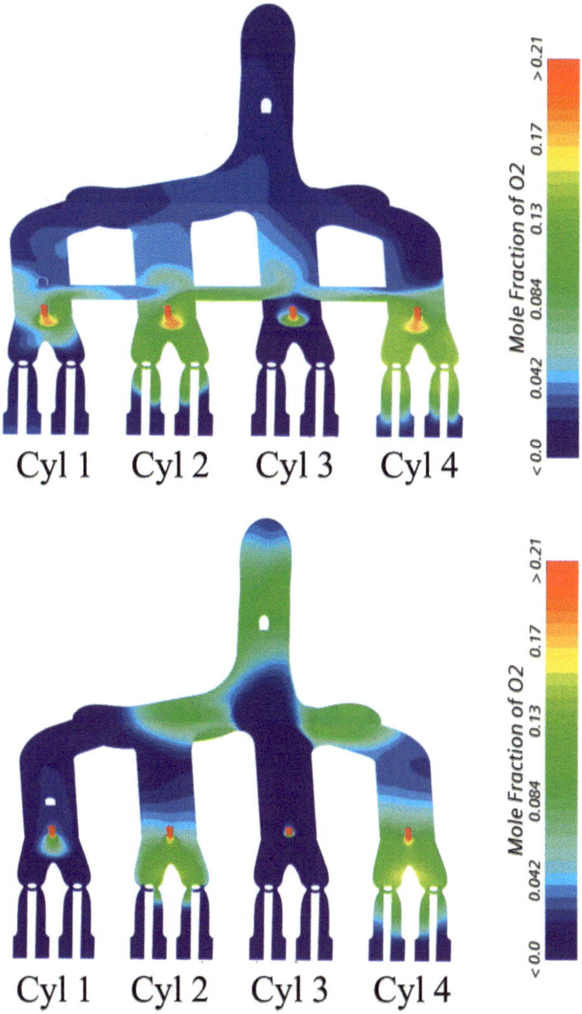

Figure 6.10: Mole fraction of O_2 during the exhaust of cyl. 3.

A comparison of exhaust temperatures was conducted using the thermocouple model at 1600 rpm, as illustrated in figure 6.11. It was observed that with the internal bypass adaptor #1 design, the inner cylinders (i.e., adaptors 2 and 3) exhibited lower temperatures compared to the conventional geometry. In contrast, the outer cylinders showed higher temperatures than those in the original design. This behaviour suggests that oxygen predominantly flows from the outer cylinder to the inner cylinder, resulting in a cooling effect.

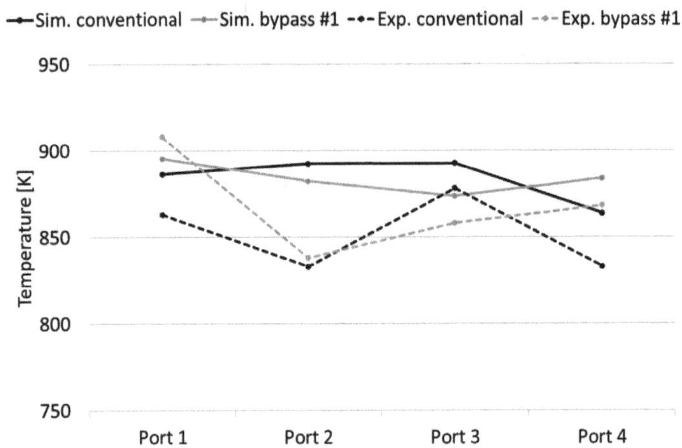

Figure 6.11: Temperature comparison between the exhaust manifolds.

Figure 6.12 shows the chemistry heat release rate and the λ_{CO} during the exhaust phase of cylinder 3 with the internal bypass adaptor #1. At this crank angle, oxygen from adaptor 4 and 2 flows back into adaptor 3, where it mixes with the exhaust gases (see figure 6.10), contributing to post-oxidation. This mixing process has been accelerated compared to the conventional design, as shown in figure 6.6. A comparison of the λ_{CO} values between figure 6.12 and figure 6.6 at the same crank angle (270 °CA a.FTDC) illustrates how the improved mixing of CO and O_2 enhances the post-oxidation process.

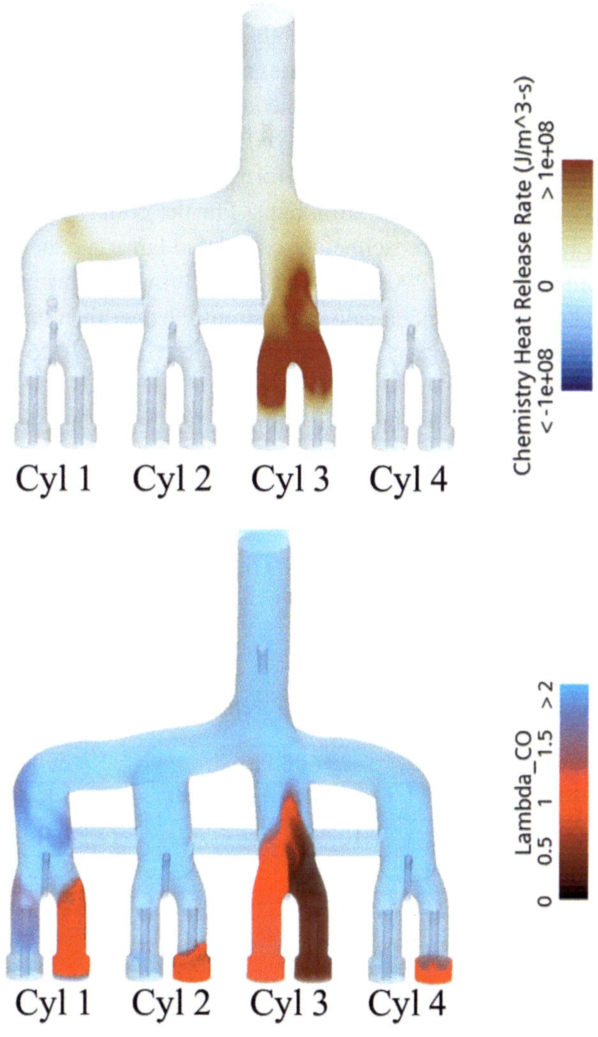

Figure 6.12: HRR and λ_{CO} during the exhaust of cyl. 3.

The different oxygen concentrations upstream of the turbine affect the measured lambda values, highlighting the importance of accurate assessment in numerical simulations. The Brettschneider formula, introduced by Brettschneider J. in [11], has been employed to track lambda. This formula calculates lambda by comparing the ratio of oxygen molecules to carbon and hydrogen molecules in the exhaust gases. Table 6.5 presents the lambda values upstream of the turbine for both engine operating points, derived from test bench measurements and calculated using the formula[11]. The formula was applied using experimental and numerical molar concentration data. For the conventional geometry, experimental lambda and Brettschneider-calculated lambda match closely at 1.10 at 1600 rpm and 1.03 at 2000 rpm. When applying the formula to numerical molar concentrations, lambda values are slightly leaner than those measured on the test bench, increasing by 2.7% at 1600 rpm (from 1.10 to 1.13) and 5.8% at 2000 rpm (from 1.03 to 1.09). A similar analysis was performed for the internal bypass adaptor #1, but only numerical data is available due to the lack of experimental measurements. In both operating points, the lambda values with the internal bypass adaptor are higher than those of the conventional design. Specifically, lambda increases by 2.7% at 1600 rpm (from 1.13 to 1.16) and 2.8% at 2000 rpm (from 1.09 to 1.12). This trend is attributed to the increased oxygen availability and the modified post-oxidation efficiency, leading to lower oxygen consumption.

Table 6.5: Lambda values for different design at 1600 rpm and 2000 rpm.

Variable	**1600 rpm**		**2000 rpm**	
	Conv.	**Adaptor #1**	**Conv.**	**Adaptor #1**
Lambda exp., [-]	1.10	-	1.03	-
Lambda BS exp., [-]	1.10	-	1.03	-
Lambda BS sim., [-]	1.13	1.16	1.09	1.12

While using the internal bypass adaptor #1 enhances mixing, it is crucial to consider the impact of the resulting increased cooling effect to determine if there is an actual improvement in post-oxidation. Therefore, it is important to evaluate the total energy released per cycle for both configurations. Another key parameter to consider is the post-oxidation rate (Pox-rate), which is calculated using equation eq. 6.2. This parameter compares the heat released by the oxidation reaction in the mesh to the amount of heat available from incoming CO and H_2, along with their corresponding lower heating values.

$$\text{Pox-rate} = \frac{\text{Heat released}}{m_{\text{CO,intake}} \cdot \text{LHV}_{\text{CO}} + m_{\text{H}_2\text{,intake}} \cdot \text{LHV}_{\text{H}_2}} \qquad \text{eq. 6.2}$$

Table 6.6 presents the total energy released per cycle due to oxidation reaction, the post-oxidation rate for the conventional exhaust manifold design, and the one with the internal bypass adaptor #1 for 1600 rpm and 2000 rpm operating condition. Both the total energy per cycle and the post-oxidation rate are lower with internal bypass adaptor #1 compared to conventional geometry. At 1600 rpm, the post-oxidation rate is 8.3% lower with the internal bypass adaptor #1, while at 2000 rpm, it is 5% lower. These results suggest that the higher availability of oxygen and the cooling effect may affect the post-oxidation.

Table 6.6: Pox-rate and energy released per cycle.

Configuration	**1600 rpm**		**2000 rpm**	
	J/cyc	**Pox-rate [%]**	**J/cyc**	**Pox-rate [%]**
Conv.	326	66.5	406	78.1
Adaptor #1	301	61.4	387	74.3

Due to the significant cooling effect caused by the oxygen flow between adaptors, particularly in the central cylinders, a new geometry was developed to improve mixing while minimizing heat losses and reducing this cooling effect. Consequently, internal bypass adaptor #2 was designed, first tested through numerical simulations, and then validated at the test bench.

6.2.2 Internal bypass adaptor #2

The new geometry introduces a connection between exhaust ports 1 and 2, and ports 3 and 4, as shown in Figure 6.9. The impact of this configuration on post-oxidation is summarized in Table 6.7, which presents the total energy released and the oxygen concentration upstream of the turbine at 1600 rpm and 2000 rpm. A comparison between the two bypass adaptors highlights that Adaptor #2 leads to a higher energy release, with an increase of approximately 3% at 1600 rpm (from 301 J/cyc to 310 J/cyc) and 2.3% at 2000 rpm (from 387 J/cyc to 396 J/cyc). This improvement can be attributed to enhanced post-oxidation, facilitated by higher temperatures associated with this design. Furthermore, the oxygen concentration upstream of the turbine is consistently lower with Adaptor #2, decreasing by 5.6% at 1600 rpm (from 3.6% to 3.4%) and 7.4% at 2000 rpm (from 2.7% to 2.5%). This reduction indicates a more efficient oxygen consumption due to improved post-oxidation reactions, reinforcing the conclusion that the revised geometry enhances the oxidation process.

Table 6.7: O_2 mol fraction for adaptor #1 and #2.

Configuration	1600 rpm		2000 rpm	
	J/cyc	Mol fr. O_2 TCup	J/cyc	Mol fr. O_2 TCup
Adaptor #1	301	3.6%	387	2.7%
Adaptor #2	310	3.4%	396	2.5%

Referring to figure 6.13, the experimental and numerical temperatures for the two analysed designs are reported. The test bench confirmed the experimental results, showing the same temperature trends in both cases and generally indicating higher temperatures with internal bypass adaptor #2.

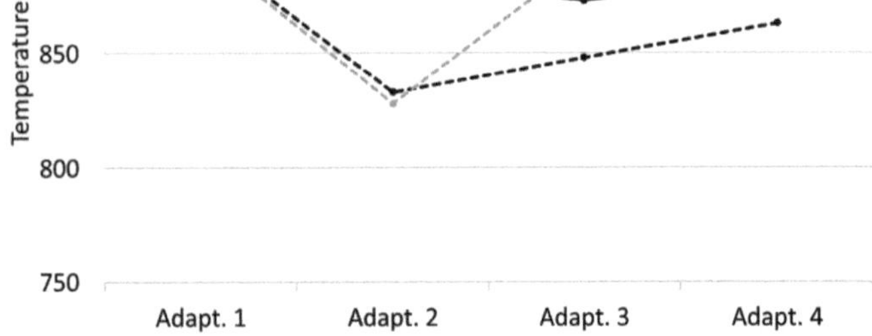

Figure 6.13: Temeprature comparison for adaptor #1 and #2.

The results obtained with the conventional exhaust manifold design and various internal bypass adaptors show a complex interaction between oxygen concentration, cooling effects, and post-oxidation efficiency. The bypass adaptor #1 resulted in an increased oxygen concentration, which improved mixing but also introduced a cooling effect that reduced post-oxidation efficiency, as evidenced by lower oxidation rates compared to the conventional manifold. The lower temperatures in the inner cylinders and the higher oxygen concentration suggested that the oxygen flow from the outer to the inner cylinders have negatively impacted the oxidation process, reducing energy released per cycle and the post-oxidation rate. Thus, the bypass adaptor #2 was introduced and it showed an improvement in exhaust gas temperature and energy released per cycle, indicating better post-oxidation. The lower oxygen concentration measured upstream of the turbine suggested greater oxygen consumption due to more effective oxidation reactions and less heat transfer through the wall. This better thermal performance and post-oxidation efficiency indicated that the new adapter design was able to better balance mixing while reducing the cooling effect observed with the previous design.

6.3 Injection strategies for enhanced post-oxidation efficiency

The analysis using the detailed exhaust model was also conducted in the context of post-injection. The boundary conditions for this model were derived from simulations performed with the 3D-CFD tool *QuickSim*, as described in Chapter 4. To account for unburned hydrocarbon oxidation, the chemical reaction mechanism previously employed ($Polimi\_1412\_H_2/CO$) needs to be updated with a more detailed mechanism, as illustrated in chapter 3.2. Consequently, this updated reaction mechanism will require a longer computation time. Post-injection involves a primary direct injection during the intake stroke, while the secondary injection is set to take place during the compression, power, or exhaust stroke. The timing of the post-injection is a critical factor in post-oxidation, directly affecting emissions. If secondary fuel injection is too late in the exhaust stroke, the post-oxidation process may not initiate effectively due to the lack of energy or time for chain-branching reactions. In the context of this work, two cases of secondary fuel injection were considered at 1600 rpm

150 Nm: secondary injection at 710 °CA (10 °CA b.FTDC), also known as late-injection, and secondary injection during the exhaust phase at crank angle 880 °CA (160 °CA a.FTDC), also known as post-injection. The selection of these two operating points is based on the experience gained from the test bench, which identified these critical points for further analysis within the virtual environment. When analysing a new post-oxidation strategy, it is essential to consider four factors that determine the effectiveness of post-oxidation: temperature, time, mixing conditions, and the availability of oxygen along with rich combustion products. Of these, time is the most critical factor in the case of post-injection. The fuel injected before the exhaust valves opening requires sufficient time and energy to initiate the oxidation process. If these conditions are not satisfied, there will be a high concentration of HC in the exhaust which will not be able to oxidise. On the other hand, fuel that is injected before the ignition point has already been crossed by the flame front, producing reactive rich combustion products that facilitate oxidation in the exhaust. In the case of late-injection, Table 6.8 displays the main concentration levels observed in both the test bench and the simulation. The results indicate that both the simulation and the test bench show similar trends. However, there are some slight differences noted in the concentrations CO, which subsequently impact the levels of CO_2. These discrepancies may be attributed to several factors, including the precision of the numerical models, the accuracy of the input parameters, and variability in the test bench conditions.

Table 6.8: Composition at adaptor 1 and turbine inlet.

Mol fr. (mol/mol)%	Adaptor 1		Turbine inlet	
	Exp.	**Sim.**	**Exp.**	**Sim.**
O_2	4.00	4.55	2.00	1.86
CO_2	10.00	8.33	12.20	10.11
CO	4.50	6.11	3.00	5.83

Figure 6.14 illustrates the heat release rate graph for late-injection (solid line) and post-injection (dashed line). In the case of late-injection (710 °CA, 10 °CA b.FTDC), significant post-oxidation occurs, which varies from cylinder to cylinder depending on the mixing conditions and lambda value of each

cylinder. For post-injection (880 °CA, 160 °CA a.FTDC), the heat released from post-oxidation is nearly zero.

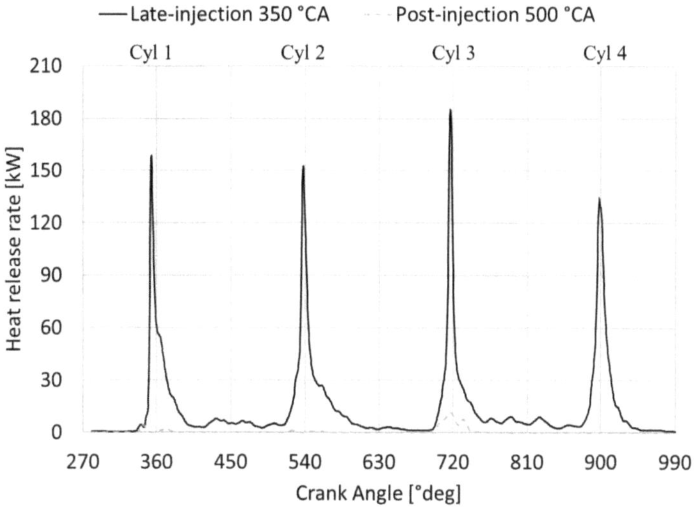

Figure 6.14: HRR for late-injection and post-injection (x-axis refers to cyl. 1).

Additionally, temperatures can be compared between the test bench and the simulation for late-injection scenarios, as illustrated in Figure 6.15, at adaptor 1 and upstream of the turbine. The trends are similar, confirming the impact of post-oxidation. At adaptor 1, the experimental temperature is 930 K, while the simulation predicts 915 K, showing a deviation of 2.2%. Upstream of the turbine, the measured temperature reaches 1150 K, with the simulation closely matching at 1140 K (a 0.9% difference). The increase in temperature from adaptor 1 to the turbine inlet further supports the presence of post-oxidation effects. The thermocouple model used in the simulation effectively captures the thermal behaviour of the exhaust system, demonstrating good agreement with experimental data.

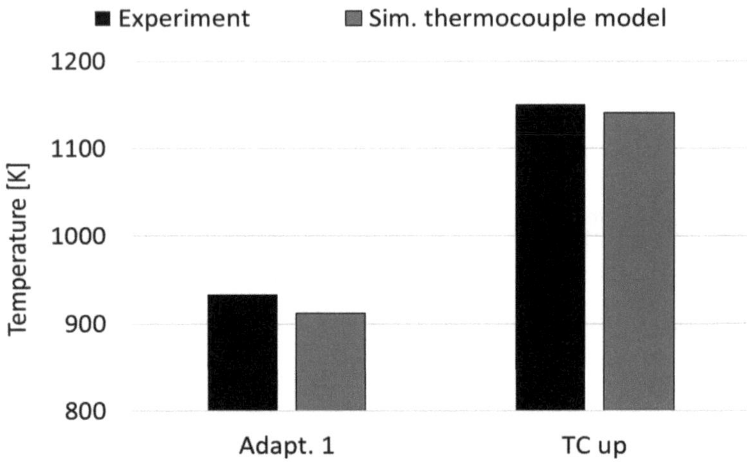

Figure 6.15: Temperature at adaptor one and before the turbine.

The results highlight that late-injection promotes post-oxidation due to the presence of more reactive combustion products, which require less energy to oxidize. In contrast, post-injection introduces fuel that must overcome a higher energy barrier to initiate oxidation, leading to incomplete post-oxidation. The temperature trends indicate a significant rise before the turbine, suggesting that the TWC will heat up more quickly, enhancing its efficiency during the initial operation phases.

7 Conclusion and Outlook

Post-oxidation has been shown to be an effective strategy for raising the temperature upstream of the TWC. This increase reduces the time needed by the TWC to reach its light-off temperature, enabling it to operate within its optimal conversion range. Several studies have explored different strategies to minimize the TWC light-off time, including active thermal management through external devices and various in-cylinder or engine control strategies. However, these methods often introduce additional system complexity or lead to the TWC operating outside of stoichiometric conditions, which can compromise its overall efficiency.

This work aims to investigate advanced post-oxidation concepts that have not been extensively addressed in the literature, with a particular focus on secondary air injection into the exhaust manifold. This technique enhances the oxidation of unburned fuel, accelerating catalyst warm-up and contributing to a reduction of pollutants emission. Once a baseline for post-oxidation strategies was established, various approaches were explored to improve the overall efficiency of the process. Different exhaust manifold designs and secondary in-cylinder injection strategies were among the potential improvements under investigation.

The effectiveness of post-oxidation is significantly influenced by mixing conditions, local temperature distributions, reaction time, and availability of oxygen and unburned combustion products. Given the challenges related to experimentally visualizing these aspects, a virtual environment is essential for studying the evolution of post-oxidation processes within the exhaust manifold. The 3D-CFD environment provide a detailed representation of flow fields and chemical reactions, enabling the quantification of mixing behaviour and the heat release rate. However, exhaust system simulations often face inaccuracies due to unreliable boundary condition. Hence, test benches typically lack the capability to provide crank-angle-resolved exhaust gas composition and spatial distribution data at the exhaust runners, which are crucial for accurately defining simulation inputs. To address these challenges, this study employed a two-step numerical approach:

M. Pipolo, *3D-CFD Virtual Development of Post-Oxidation Strategies with Secondary Air Injection for Advanced Internal Combustion Engines*,
Wissenschaftliche Reihe Fahrzeugtechnik Universität Stuttgart,
https://doi.org/10.1007/978-3-658-51343-6_7

1. A full-engine 3D-CFD model was used to investigate mixture formation, cylinder-to-cylinder imbalance, and the post-oxidation precursors.
2. A detailed 3D-CFD exhaust manifold model incorporated a refined chemical reaction mechanism to track pollutant formation and assess oxidation efficiency.

Boundary conditions for the detailed exhaust model were extracted from the full engine simulations, integrating time-resolved exhaust gas compositions and distribution to enhance accuracy. The numerical results were validated against experimental data measured by Chiba University, confirming the importance of the collaboration between the two institutions in achieving a clear and comprehensive understanding of the research topic. The refined exhaust model proved to be an efficient tool, significantly reducing computational costs while enabling high-fidelity investigations of post-oxidation phenomena. Following the validation phase, the study first established a reference for the post-oxidation occurring at different engine operating conditions with secondary air injection. Once this baseline was defined, further optimization strategies were explored, including increasing the temperature of the injected secondary air, developing optimized exhaust manifold geometries, and introducing direct in-cylinder secondary fuel injection.

7.1 Insights from the 3D-CFD full engine simulation

Complete engine simulations provide essential insights for predicting the key phenomena that characterize the engine operating points and their post-oxidation processes. This investigation specifically focuses on mixture formation and inhomogeneities among cylinders. Understanding the parameters that trigger post-oxidation phenomena is crucial, as it forms the foundation for examining the post-oxidation process itself. During SAI and no-SAI operations, inhomogeneities within the cylinder create distinct zones with significantly rich conditions due to incomplete mixture formation. During combustion, the flame front crosses this areas, producing a high concentration of rich combustion products. Additionally, these inhomogeneities lead to slight variations in the distribution of unburned gases across the exhaust valves. This causes differences in the amounts of reactive species released through each valve, subsequently affecting the post-oxidation process in the exhaust manifold. Furthermore, when the engine operates without SAI, it runs with an excess of air in the cylinder. Despite this leaner operation, the mixture is still not completely homogenized. Therefore, areas with rich conditions in the combustion chamber remain, producing CO, H_2, and HC. These reactive products can undergo oxidation in the exhaust system if there is enough oxygen. However, if oxygen is lacking, the oxidation process cannot take place.

The imbalance among cylinders has also been assessed, potentially caused by factors such as engine design, the location and geometry of intake and exhaust ports, and fuel delivery systems. By analysing in-cylinder mass and air-fuel ratio values at different speeds, it is possible to determine which cylinders take in more air and how this affects the combustion process. Typically, the outer cylinders take in more air and, therefore, run in leaner condition compared to the inner cylinders.

Another post-oxidation strategy explored in the project involves secondary fuel injection within the cylinder and SAI in the exhaust manifold. Primary direct injection occurs during the intake stroke, while secondary injection is planned for either the compression or exhaust stroke. Two cases of secondary fuel injection were considered at 1600 rpm and 150 Nm: secondary injection at 710 °CA (10 °CA before TDC), known as late-injection, and secondary injection during the exhaust phase at a crank angle of 880 °CA (160 °CA after TDC), known as post-injection. The engine operates with a rich in-cylinder lambda in the late-injection case. Hence, the flame front crosses areas where the oxygen is not sufficient completely burn the fuel, resulting in the formation of highly reactive incomplete combustion products. These products later react in the exhaust system when interacting with oxygen coming from the SAI. In the post-injection case, fuel evaporation occurs shortly after the exhaust valves opening, causing a temperature drop. This reduction in temperature negatively affects post-oxidation efficiency in the exhaust manifold.

7.2 Post-oxidation strategies and optimization

A key advantage of using a refined exhaust manifold model for post-oxidation studies is its ability to integrate a chemical reaction mechanism while maintaining reasonable computational time. This allows for the evaluation of multiple post-oxidation strategies within a reduced time frame. Once it was determined that secondary air injection and a rich lambda engine operations offered benefits over the same engine operating point without SAI, the optimization of post-oxidation strategies was explored as follows:

1. Increasing the temperature of secondary air injection: higher temperatures for SAI enhanced oxidation efficiency by accelerating reaction kinetics. However, the SAI temperature must remain consistent with realistic values found in conventional applications.
2. New concepts of the exhaust manifold: these designs features a 20 mm diameter internal hole on the side of each exhaust line. This hole allows exhaust gases from the exhaust port to flow into other exhaust lines. Internal bypass adaptor #1 bypasses all exhaust lines, while internal bypass adaptor #2 separately bypasses exhaust ports 1 and 2, as well as exhaust ports 3 and 4. The adaptor #1 was designed to improve mixing by recirculating exhaust gases before the turbine, thereby increasing interactions between fresh oxygen and unburned species. However, localized cooling effects and higher heat losses negatively impacted oxidation efficiency. The adaptor #2 reduced the heat losses and the cooling effect above-mentioned. Therefore, more uniform mixing was achieved while keeping sufficient high exhaust gas temperatures, improving oxidation efficiency.
3. Post-injection strategies: secondary fuel injection before the ignition point resulted in a highly reactive fuel-air mixture, maximizing oxidation potential and heat release in the exhaust. Fuel injection during the exhaust stroke proved less effective due to the lower reactivity of unburned species. Additionally, the decrease in temperature caused by the evaporation of fuel before the exhaust valve opening reduced overall post-oxidation efficiency and limited heat generation.

The study developed a comprehensive framework for assessing post-oxidation efficiency across various strategies through this numerical approach. Detailed chemical kinetics and high-resolution analysis provided deeper insights into energy released by oxidation reactions, ensuring a solid basis for optimizing post-oxidation performance.

7.3 Outlook: the 3D-CFD turbocharger model

This work successfully demonstrated that post-oxidation through secondary air injection is an effective strategy for increasing exhaust temperatures upstream of the TWC. The findings highlight how advanced numerical tools can accelerate the development of optimized post-oxidation strategies, offering a clear pathway toward improved emissions control and increased engine efficiency. The growing computational power and the availability of more detailed numerical models allow for the investigation of phenomena that were previously not considered, such as incorporating a 3D turbocharger model as an integral part of 3D-CFD simulation. Introducing the turbocharger into the virtual engine test bench enables the application of boundary conditions that match the real working conditions, which is particularly valuable in the early stages of engine development when no experimental data are available for calibrating and validating 1D and 3D simulation models. Simplified approaches were used to balance computational time with the reliability of results [34]. Among these, the following are of particular importance:

- 0D Turbocharger Model: This model enables continuous data exchange between the 3D simulation and the 0D turbocharger model, with processed information such as a new shaft speed, returned as global data in the form of an internal boundary condition. This approach ensures that the boundary conditions for full engine simulations are based on air at ambient pressure and temperature, without the need for specific 3D flow assumptions, thereby reducing dependency on boundary conditions.
- 3D Turbocharger Model: Figure 7.1 shows the 3D turbocharger model implemented to the full engine domain. The boundary conditions at the intake are directly taken from the compressor outlet, while a turbine geometry is implemented at the exhaust. This allows for a high level of detail on the

exhaust side, which is crucial for accurately evaluating the flow field and studying local mixing phenomena.

The second approach is only possible if the simulation time remains acceptable and the model is stable, given the many pressure pulsations present at the exhaust of a turbocharged engine that must be accurately represented. The implementation of the 3D turbocharger model provided initial insights into how the turbine influences the flow field in the exhaust manifold, particularly in relation to post-oxidation processes. However, this remains a preliminary step, as the model has not yet undergone validation. Future studies should focus on refining this model through dedicated experimental or high-fidelity numerical validation to assess its accuracy and predictive capabilities.

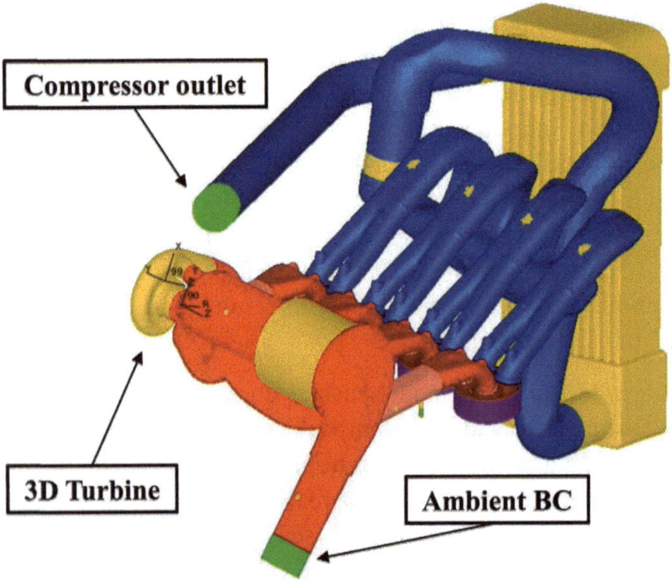

Figure 7.1: Full engine and 3D turbocharger model.

To illustrate the main differences between the two approaches, figure 7.2 shows the velocity field of the 0D turbocharger (on the right) and the 3D turbocharger model (on the left). Similar observations can be made by examining figure 7.3, which illustrates the exhaust of cylinder 4.

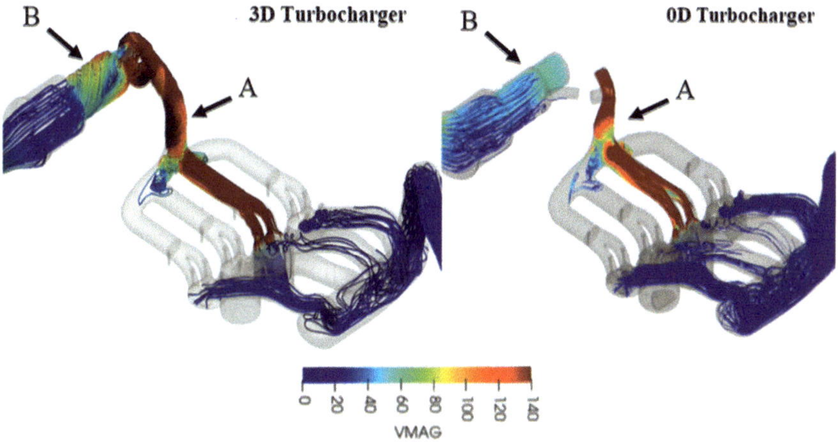

Figure 7.2: Full engine and 3D turbocharger model during exhaust cyl. 3.

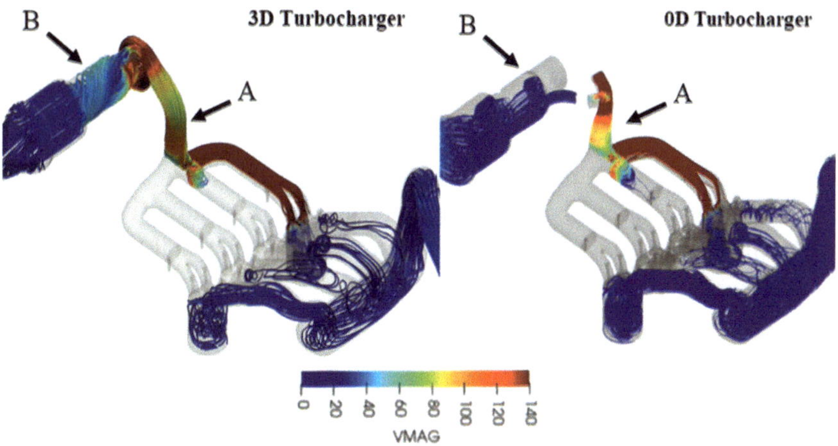

Figure 7.3: Full engine and 3D turbocharger model during exhaust cyl. 4.

These specific moments highlight two distinct positions:

- Position A: Upstream of the turbine, the 3D turbine provides a more accurate representation of the mixing conditions compared to the 0D model. The latter is influenced by the 1D mass flow boundary at the turbine inlet, which affects the flow field. In contrast, the 3D turbocharger allows for the investigation of local mixing phenomena that may lead to post-oxidation reactions. As a result, since no 1D boundary is present, the turbine influences the flow field within the exhaust manifold.
- Position B: The flow in the 0D turbocharger model appears linear at the turbine outlet because the implemented mass flow is one-dimensional. In the 3D model, however, the turbine operation is fully represented, ensuring that no flow field information is lost.

Integrating the turbocharger model enables a detailed energy balance analysis of the entire engine system under both steady-state and transient operating conditions. This means that any change in the operating state of the internal combustion engine will result in a corresponding change in the turbocharger operating state, and vice versa. This coupling serves as a crucial point for conducting transient load simulations across hundreds of operating cycles. As a result, the application of 3D-CFD is extended into areas that were previously only accessible through test bench experiments and 1D simulations. This approach provides valuable insights into complex phenomena such as turbo lag, turbo matching, and emission formation during transient load changes.

The findings of this study contribute to the ongoing advancement of cleaner and advanced internal combustion engine technologies. Future research should focus on refining injection strategies, optimizing exhaust manifold designs, and integrating advanced numerical simulations with experimental validation to maximize post-oxidation efficiency.

Bibliography

[1] European council, clean and sustainable mobility report (https://www.consilium.europa.eu/).

[2] Fit for 55: Delivering the eu's 2030 climate target on the way to climate neutrality (https://ec.europa.euft/).

[3] Paris agreement (https://eur-lex.europa.eu/), 10/2016.

[4] A. Schmid, M. Grill and M. Boettcher. Development of a quasi-dimensional combustion model for stratified si engines. Technical report, SAE technical paper, 2009.

[5] M. Albrecht, H.-P. Deeg, D. Schwarzenthal, and P. Eilts. The influence of fuel composition and renewable fuel components on the emissions of a gdi engine. Technical report, SAE Technical Paper, 2020.

[6] J. Anderson. Computational fluid dynamics. 1995.

[7] M. Bargende. Berechnung und analyse innermotorischer vorgange. lecture notes. 2013.

[8] M. Battistoni, J. Zembi, D. Casadei, F. Ricci, R. Martinelli, C. Grimaldi, M. Sana, M. Brignone, A. Mantovanelli, and E. Milani. Burner development for light-off speed-up of aftertreatment systems in gasoline si engines. 06 2022.

[9] J. Blazek. *Computational fluid dynamics: principles and applications*. Butterworth-Heinemann, 2015.

[10] M. Borland and F. Zhao. Application of secondary air injection for simultaneously reducing converter-in emissions and improving catalyst light-off performance. Technical report, SAE technical paper, 2002.

[11] J. Brettschneider. Berechnung des luftverhältnisses lambda von luft-kraftstoff-gemischen und des einflusses von messfehlern auf lambda. *Bosch Tech Ber*, 6(4), 1979.

M. Pipolo, *3D-CFD Virtual Development of Post-Oxidation Strategies with Secondary Air Injection for Advanced Internal Combustion Engines*, Wissenschaftliche Reihe Fahrzeugtechnik Universität Stuttgart,
https://doi.org/10.1007/978-3-658-51343-6

[12] CD adapco. Methodology - STAR-CD Version 4.20, 2013.

[13] M. Chiodi. *An Innovative 3D-CFD-Approach towards Virtual Development of Internal Combustion Engines.* PhD thesis, University of Stuttgart, 2010.

[14] A. Cuoci, A. Frassoldati, T. Faravelli, and E. Ranzi. Numerical modeling of laminar flames with detailed kinetics based on the operator-splitting method. *Energy & fuels*, 27(12):7730–7753, 2013.

[15] F. Cupo. *Modeling of real fuels and knock occurrence for an effective 3D-CFD virtual engine development.* Springer, 2021.

[16] A. S. D. Rether, M. Grill and M. Bargende. Quasi-dimensional modeling of ci-combustion with multiple pilot- and post injections. Technical report, SAE technical paper, 2010.

[17] C. Dardiotis, G. Martini, A. Marotta, and U. Manfredi. Low-temperature cold-start gaseous emissions of late technology passenger cars. *Applied Energy*, 111:468–478, 2013.

[18] F. Del Pero, M. Delogu, and M. Pierini. Life cycle assessment in the automotive sector: A comparative case study of internal combustion engine (ice) and electric car. *Procedia Structural Integrity*, 12:521–537, 2018.

[19] J. Demuynck, P. M. Villafuerte, D. Bosteels, A. Kuhrt, M. Brauer, M. Sens, J. Williams, C. Chaillou, and V. Gordillo. Advanced emission controls and e-fuels on a gasoline car for zero-impact emissions. *SAE International Journal of Advances and Current Practices in Mobility*, 5(2022-01-1014):1063–1069, 2022.

[20] European Commission and Joint Research Centre and J. Suarez, D. Komnos et al. 2025 and 2030 co2 emission targets for light duty vehicles, 2023.

[21] European Environment Agency. Sustainability of europe's mobility systems (https://www.eea.europa.eu/), 10/2024.

[22] European Parliament, Council of the European Union. Regulation (eu) 2021/1119 (european climate law) (https://eur-lex.europa.eu/legal-content/en/txt/?uri=celex:22016a1019(01)), 2021.

[23] D. E.Winterbone and R. J. Pearson. Design techniques for engine manifolds: Wave action methods for ic engines. 1999.

[24] G. C. u. E. N. G. Comini. *Fondamenti di termofluidodinamica computazionale*. Padova: Servizi Grafici Editoriali, 2014.

[25] J. Gao, G. Tian, A. Sorniotti, A. E. Karci, and R. Di Palo. Review of thermal management of catalytic converters to decrease engine emissions during cold start and warm up. *Applied Thermal Engineering*, 147:177–187, 2019.

[26] I. Glassman, R. A. Yetter, and N. G. Glumac. *Combustion*. Academic press, 2014.

[27] C. Gong, K. Huang, B. Deng, and X. Liu. Catalyst light-off behavior of a spark-ignition lpg (liquefied petroleum gas) engine during cold start. *Energy*, 36(1):53–59, 2011.

[28] D. G. Goodwin, R. L. Speth, H. K. Moffat, and B. W. Weber. Cantera: An object-oriented software toolkit for chemical kinetics, thermodynamics, and transport processes. *Zenodo*, 2018.

[29] Gregory P. Smith, David M. Golden, Michael Frenklach NW, Moriarty BE, Mikhail Goldenberg, C. , Thomas Bowman RK, Hanson SS, William C. , et al. Gri-mech 3.0; http://www.me.berkeley.edu/gri_mech/.

[30] F. Haußmann and B. Höpke. Optimierung des dynamischen verhaltens hochaufgeladener ottomotoren nach den kriterien bester verbrauch und bestes transientes ansprechverhalten. *ATL, Frankfurt am Main*, 2015.

[31] J. Hofstetter, P. Boucharel, F. Atzler, and G. Wachtmeister. Fuel consumption and emission reduction for hybrid electric vehicles with electrically heated catalyst. *SAE International Journal of Advances and Current Practices in Mobility*, 3(2020-37-0017):702–714, 2020.

[32] L. E. Hombach, L. Doré, K. Heidgen, H. Maas, T. J. Wallington, and G. Walther. Economic and environmental assessment of current (2015) and future (2030) use of e-fuels in light-duty vehicles in germany. *Journal of cleaner production*, 207:153–162, 2019.

[33] J. Hu, H. C. Frey, and B. Y. Boroujeni. Contribution of cold starts to real-world trip emissions for light-duty gasoline vehicles. *Atmosphere*, 14(1), 2023.

[34] A. Kächele. *Turbocharger Integration into Multidimensional Engine Simulations to Enable Transient Load Cases*. Springer, 2020.

[35] C. Koehlen, E. Holder, and G. Vent. Investigation of post oxidation and its dependency on engine combustion and exhaust manifold design. *SAE Transactions*, pages 1264–1275, 2002.

[36] A. C. Kulzer, H. P. Deeg, J. Villforth, D. Schwarzenthal, M. Schilling, E. Barrientos, B. Kanach, and M. Lindner. Sustainable mobility using fuels with pathways to low emissions. *SAE International Journal of Advances and Current Practices in Mobility*, 2(2020-01-0345):1870–1892, 2020.

[37] M. Kumar, T. Kuboyama, N. Hasegawa, and Y. Moriyoshi. Scavenging phenomena based post-oxidation in exhaust manifold of a turbocharged spark ignition engine. Technical report, SAE technical paper, 2019.

[38] M. Kumar, S. Moeeni, T. Kuboyama, and Y. Moriyoshi. Performance improvement of turbocharged si engine by post-oxidation enhancement in exhaust gas in-homogeneity. *International Journal of Engine Research*, 22(9):2931–2944, 2021.

[39] M. Kumar, S. Moeeni, T. Kuboyama, Y. Moriyoshi, T. Aizawa, and M. Bargende. Analysis of post-oxidation phenomena in a turbocharged disi engine focused on gas and pm emissions and gas temperature. *International Journal of Automotive Engineering*, 12(4):116–124, 2021.

[40] M. Kumar, S. Moeeni, T. Kuboyama, Y. Moriyoshi, J. Przewlocki, R. Tromellini, M. Grill, M. Chiodi, and M. Bargende. Investigation of h2 formation characterization and its contribution to post-oxidation phenomenon in a turbocharged disi engine. Technical report, SAE technical paper, 2020.

[41] C. K. Law. *Combustion physics*. Cambridge university press, 2010.

[42] D. Lee and J. B. Heywood. Effects of secondary air injection during cold start of si engines. *SAE International Journal of Engines*, 3(2):182–196, 2010.

[43] D. R. M. Grill, M. Bargende and A. Schmid. Quasi-dimensional and empirical modeling of compression-ignition engine combustion and emissions. Technical report, SAE technical paper, 2010.

[44] S. McAllister, J.-Y. Chen, and A. C. Fernandez-Pello. *Fundamentals of combustion processes*, volume 304. Springer, 2011.

[45] G. P. Merker and R. Teichmann. Grundlagen verbrennungsmotoren. 2014.

[46] A. Mokros, P. Demel, F. Knost, M. Münz, and C. Beidl. Synthetic, regenerative fuels (refuels) as enabler for climate neutral mobility and transport. In *Internationaler Motorenkongress 2020*, pages 399–403. Springer, 2020.

[47] Council of the European Union. 'fit for 55': Council adopts regulation on co2 emissions for new cars and vans. (https://www.consilium.europa.eu/)

[48] I. C. on Clean Transporation. Effects of battery manufacturing on electric vehicle life-cycle greenhouse gas emissions (https://theicct.org/)

[49] A. Onorati, G. Ferrari, and G. D'Errico. Secondary air injection in the exhaust after-treatment system of si engines: 1d fluid dynamic modeling and experimental investigation. *SAE transactions*, pages 544–555, 2003.

[50] B. Pathak, A. Shekh, N. Patel, and V. Patel. Review analysis on cfd techniques and numerical methods for ic engines fueled with diesel and biofuels. *Journal of Advanced Research in Fluid Mechanics and Thermal Sciences*, 110(1):40–62, 2023.

[51] M. Pelucchi, C. Cavallotti, T. Faravelli, and S. Klippenstein. H-abstraction reactions by oh, ho 2, o, o 2 and benzyl radical addition to o 2 and their implications for kinetic modelling of toluene oxidation. *Physical Chemistry Chemical Physics*, 20(16):10607–10627, 2018.

[52] L. Postrioti, M. Battistoni, J. Zembi, G. Brizi, M. Sana, M. Brignone, F. Napoli, S. Pizza, and E. Milani. Experimental and numerical analysis of a swirled fuel atomizer for an aftertreatment diesel burner. 08 2023.

[53] J. Pritchard and W. K. Cheng. Effects of secondary air on the exhaust oxidation of particulate matters. *SAE International Journal of Engines*, 8(3):1088–1097, 2015.

[54] J. Przewlocki, R. Tromellini, M. Grill, M. Chiodi, and M. Bargende. Investigation and 1d modelling approach on scavenging air post-oxidation inside the exhaust manifold of a disi engine. Technical report, SAE technical paper, 2021.

[55] PWC Deutschland. Energiewende outlook: Transportation sector. (https://store.pwc.de/)

[56] K. M. Reiter, Matthew S.and Kockelman. The problem of cold starts: A closer look at mobile source emissions levels. *Transportation Research Part D: Transport and Environment*, 43:123–132, 2016.

[57] D. Serrano, J. Lavy, A. Kleemann, S. Zinola, J.-P. Dumas, S. Le Mirronet, and D. Heitz. Post oxidation study during secondary exhaust air injection for fast catalyst light-off. Technical report, SAE technical paper, 2009.

[58] Siemens. StarCCM+ User Guide, 2013.

[59] D. K. Srivastava, A. K. Agarwal, A. Datta, and R. K. Maurya. *Advances in internal combustion engine research.* Springer, 2018.

[60] G. Stiesch. Elektronisches management motorischer fahrzeugantriebe. pages 88–102, 2010.

[61] U. Tietge, J. Dornoff, and P. Mock. Co2 emissions from new passenger cars in europe: Car manufacturers' performance in 2023. 09 2024.

[62] R. Tromellini, M. Kumar, S. Moeeni, M. Chiodi, M. Bargende, T. Kuboyama, and Y. Moriyoshi. 3d-cfd full engine simulation application for post-oxidation description. Technical report, SAE Technical Paper, 2021.

[63] R. Tromellini, J. Przewlocki, F. Cupo, M. Bargende, and M. Chiodi. Analysis of scavenging air post-oxidation by means of 3d-cfd simulation including reaction mechanism. In *20. Internationales Stuttgarter Symposium: Automobil-und Motorentechnik*, pages 421–437. Springer, 2020.

[64] S. R. Turns et al. *Introduction to combustion*, volume 287. McGraw-Hill Companies New York, NY, USA, 1996.

[65] H. K. Versteeg. *An introduction to computational fluid dynamics the finite volume method, 2/E*. Pearson Education India, 2007.

[66] K. Wagemann and F. Ausfelder. *E-Fuels-mehr als eine Option*. DECHEMA Gesellschaft für Chemische Technik und Biotechnologie eV Frankfurt/Main, 2017.

[67] P. Weber, J. Buchgeister, O. Toedter, and T. Koch. Life cycle assessment as a tool for analyzing the co 2 footprint of passenger cars with different powertrains. In *Internationaler Motorenkongress 2020*, pages 387–398. Springer, 2020.

[68] M. Weilenmann, J.-Y. Favez, and R. Alvarez. Cold-start emissions of modern passenger cars at different low ambient temperatures and their evolution over vehicle legislation categories. *Atmospheric Environment*, 43(15):2419–2429, 2009.

[69] M. Weiss, E. Paffumi, M. Clairotte, Y. Drossinos, T. Vlachos, P. Bonnel, and B. Giechaskiel. Including cold-start emissions in the real-driving emissions (rde) test procedure. 01 2017.

[70] J. Yadav, V. Betgeri, B. Graziano, A. Dhongde, B. Heuser, M. Schönen, and N. Sittinger. Renewable drop-in fuels as an immediate measure to reduce co 2 emissions of heavy-duty applications. In *Internationaler Motorenkongress 2020*, pages 353–372. Springer, 2020.

MIX
Papier aus verantwortungsvollen Quellen
Paper from responsible sources
FSC® C105338

If you have any concerns about our products,
you can contact us on
ProductSafety@springernature.com

In case Publisher is established outside the EU,
the EU authorized representative is:
Springer Nature Customer Service Center GmbH
Europaplatz 3, 69115 Heidelberg, Germany

Printed by Libri Plureos GmbH
in Hamburg, Germany